The Institute of Biology's
Studies in Biology no. 154

Wetland Ecology

John R. Etherington

B.Sc., A.R.C.S., Ph.D., D.I.C.
Senior Lecturer in Plant Science,
University College, Cardiff

Edward Arnold

First published 1983
by Edward Arnold (Publishers) Limited
41 Bedford Square, London WC1 3DQ

British Library Cataloguing in Publication Data

Etherington, John R.
    Wetland ecology.—(The Institute of Biology's studies in biology ISSN 0537–9024; 154)
    1. Wetland ecology
    I. Title   II. Series
    574.5'26325   QH541.5.M3

    ISBN 0–7131–2865–8

Printed and bound in Great Britain at
The Camelot Press Ltd, Southampton

General Preface to the Series

Because it is no longer possible for one textbook to cover the whole field of biology while remaining sufficiently up to date, the Institute of Biology proposed this series so that teachers and students can learn about significant developments. The enthusiastic acceptance of 'Studies in Biology' shows that the books are providing authoritative views of biological topics.

The features of the series include the attention given to methods, the selected list of books for further reading and, wherever possible, suggestions for practical work.

Readers' comments will be welcomed by the Education Officer of the Institute.

1983   Institute of Biology
                                                         20 Queensbury Place
                                                         London SW7 2YY

Preface

Wetlands, a distinctive group of habitats intermediate between aquatic and terrestrial ecosystems, have a specialized vegetation which copes with the vagaries of fluctuating watertables, the chemical oddities of anaerobic soils and, in some cases, the problems of inundation with saline water. Extensive areas are often inaccessible to large predatory mammals including man and so provide a refuge for a diversity of beautiful and unusual wildlife.

Because decomposition is slow in waterlogged soils, wetlands also preserve a stratified record of past succession, pollen rain and occasionally archaeological artefacts hidden in the peats and sediment layers below the present surface, a record which is as much at risk from drainage as the living community above.

Tropical and subtropical wetlands have been exploited and created to produce rice which feeds more than half the human population. It is a remarkable crop, many varieties growing in an agricultural reedswamp with anaerobic soils so reducing that they contain toxic amounts of various metals and sulphides under an atmosphere which is rich in methane and hydrogen. In this unlikely environment rice is a productive crop and probably the most waterlogging tolerant of all plants.

This book is intended to introduce the fascinating complex of adaptations by which plants, animals and micro-organisms live, thrive and interact within these diverse but frighteningly vulnerable communities.

Cardiff, 1983   J.R.E.

Contents

1 Wetlands: from tropics to tundra

1.1 Freshwater wetlands

Compared with the vegetation of well-drained soils, wetlands have a world-wide similarity which over-rides climate and is imposed by the common characteristics of a free water supply and the abnormally hostile chemical environment which plant roots must endure. The component species have morphological, anatomical and physiological adaptations which allow them to cope with intermittent flooding, lack of oxygen (anoxia) and the consequent chemical reduction of the soil. Constant flooding limits accessibility to many large herbivores and, particularly in temperate climates, reedswamps and fens are consequently less moulded by grazing-pressure than the adjacent dry-land vegetation.

Aquatic marginal habitats are associated with the static water of lakes and ponds, with flowing water in riverside situations and with the rather special diurnal oscillations of waterlevel in freshwater tidelands above the reach of saline water. The most widespread vegetation is reedswamp of which the emergent species belong predominately to families of monocotyledons such as the Cyperaceae, Gramineae and Sparganiaceae. In some parts of the world they are of enormous extent: over 150 000 km^2 in the Sudd of the upper Nile and more than 10 000 km^2 in the Botswana Okovanggo. In Europe the great reedswamp tract of the Camargue has largely been drained for rice and other crop production.

Slow-flowing rivers and streams are fringed by a reedswamp which is more or less indistinguishable from that of still waters but faster-flowing headwaters and immature rivers of unpredictable catchments often lack any specialized riverine vegetation.

Classifications are artificial and it is merely a matter of convenience to separate *terrestrial wetlands* from those adjacent to water-bodies in which the watertable is essentially controlled by the behaviour of the open-water. Terrestrial wetland soils may be mainly mineral and carry a *marshland* vegetation with its roots in waterlogged soil but suffering only occasional flooding. Such sites fail to accumulate an organic *peat* covering because they are relatively well oxygenated and of moderate to high mineral nutrient status so that microbial respiratory oxidation is not impeded. The vegetation of temperate marshland ranges from species-rich herbaceous cover, scrub woodland of willow and alder (*Salix* spp. and *Alnus* spp.), if the groundwater is well provided with nutrients, to a *wet heathland* species-poor association in oligotrophic conditions.

The peatlands or *mires* of the world contrast with these soils in the sometimes enormous accumulations of semi-decomposed plant remains which fail to oxidize because of anoxia or oligotrophy. The peat covering acts as a giant sponge, carrying a perched watertable upward, in pace with its growth. A variety of mire types form in response either to accumulation of drainage water over impermeable rocks, glacial drift and permafrost, or to wetness in almost any geological environment, caused by a very large precipitation to evaporation ratio (P/E). Nutrient-rich groundwater mires (*valley mires*) may carry a species-rich *fen* vegetation which intergrades with the acid *wet-heath* of oligotrophic groundwater mires while rainfed peatlands (*blanket mires* and *raised mires*) are inevitable acid, mineral deficient and vegetated with species-poor wet heathland in which bogmosses (*Sphagnum* spp.) play a prominent part (Plate 1). In the tropics, peats may carry a reedswamp-fen, riverine *swamp forest* or, in the case of rainfed peat, a species-poor swamp forest. The interrelationships of these ecosystem types are summarized in Fig. 1-1.

1.2 Agricultural wetlands

More than half of the world's population is supported by rice (*Oryza sativa*), a domesticated tropical reedswamp grass. A large portion of its production is based on *paddy* culture in which the young seedlings are planted into shallow water, flooding a previously cultivated, usually green-manured soil. Its varieties range from deep-water 'floating' rice to upland rices which are never flooded. The wetland varieties are probably the most waterlogging tolerant of all plants, their warm, organic-rich, microbially-active soils producing extremes of anoxia and chemical reduction.

In parts of South America, notably the Valley of Mexico, the Spanish conquistadors found Aztec *chinampa* or 'floating-garden' agriculture in the swampy lake-shore areas. Sediment and aquatic plants were dredged to make peaty beds of which the sides were reinforced by woven branches and planted willows. Maize, beans, peppers and grain-amaranths were grown to which, today, have been added a variety of truck-crops, for example onions and carrots. The name floating-garden was given because seedlings were raised on floating mats of vegetation and peat which could be towed to the planting site and transferred to the chinampa in peat blocks. A similar agriculture is practised in the Vale of Kashmir in the Himalayas.

In parts of lowland Britain and north Europe *water-meadows* are managed as grazing and hay-producing grassland. Sluices, ditches and embankments are used to flood the land with nutrient-rich silt-laden water when the rivers are in spate. During the spring months this raises soil temperature, promotes nitrogen-fixation and stimulates an early herbage crop. Few such areas now survive modern mechanized agriculture but where they persist, their species-rich grassland flora is further supplemented by many marshland and other wetland species to form a vegetation type of great interest and conservational importance.

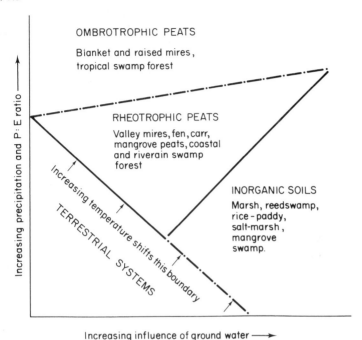

Fig. 1-1   The interrelated effects of groundwater and precipitation: evaporation ratio on soil-formation. The continuous line marks an arbitrary boundary between organic and inorganic soils.

1.3 Maritime saline wetlands

The great *salt marshes* of Europe and North America, fringing the tidal flats where the sea never rests in its ceaseless cycle of accretion and erosion give one of the most vivid insights of succession and degeneration. Like many other waterside vegetations a centimetre or two difference in elevation is ecologically determinate, not only in this case because of the duration of waterlogging but also in relation to inundation with saltwater. These habitats are dominated by halophytic herbs and dwarf shrubs many of which are strongly succulent and some others provided with salt-excreting glands or epidermal hairs. The families Chenopodiaceae, Compositeae and Gramineae are particularly well represented amongst salt-marsh halophytes.

Within the equatorial belt bounded by latitudes 30⁰N and 30⁰S the salt marshes generally give way to a *mangrove swamp* vegetation of small trees which flourish in tidal or brackish sediments and similarly share a composite physiological and morphological adaptation to both waterlogging and salinity (Fig. 1-2). In many cases, freely flushed by tidal channels and constantly dropping litter of their evergreen leaves, mangrove swamps export large amounts of organic matter to nearshore food chains.

Fig. 1-2 (**a**) Mangrove swamp of *Rhizophora mangle* with prominent stilt-roots. Atlantic coast of Mexico. (**b**) Stilt roots of black mangrove (*Rhizophora mangle*) and emergent breathing roots (pneumatophores) of *Avicennia germinans*. Very large lenticels are visible on both structures.

Neither sea nor land, the salt marshes and mangroves are tidal interfaces between shallow waters and the coastal catchment which benefit nutritionally, in soil aeration and consequent high productivity from the tidal pulsing. In addition to their exports of litter they internally support large numbers of molluscs, worms, crabs, shrimps and fish. The mudflats of the tidal salt marshes scurry and cry with myriad wading birds, dependant on this production, while the outer fringes of many mangroves are literally weighed down with bird life which feeds in the shallow waters.

Estuaries, throughout the world, are outstandingly productive, particularly of planktonic algae and their dependent food-chains which are at the receiving end of terrestrial outputs: dissolved nutrients and organic materials together with particulate litter. Tidal mudflats fringing both mangroves and salt marshes support great beds of the few families of marine angiosperms: the Zosteraceae and Cymodocaceae (sea grasses) are the most prominent. *Zostera marina* (eel grass) is frequent in northern waters and several genera of the Cymodocaceae such as *Cymodocea* and *Thallassia* in tropical waters.

1.4 Inland salines

The arid zones of the world provide an atmospheric sink of water vapour so strong that cyclic supply cannot keep pace and evaporation is either limited, or includes a component of rising capillary groundwater which carries dissolved solutes to the suface where they become so enriched that *salt desert* habitats are formed. Many of these soils have waterlogged characteristics because a high watertable promotes the capillary contribution to evaporation, and once salt has accumulated to the extent of forming surface efflorescences in the typical *solonchak* soils, the reduced water-vapour pressure allows the surface soils to remain permanently wet and inhabited by halophilic and often anaerobic bacteria. Accidental raising of watertables by irrigation has greatly increased the areas of such soils in the arid zones.

Just as the gleyed salt-rich soils of these deserts share characteristics with those of the coast, so are their plant inhabitants related. The predominance of Chenopodiaceae in the halophyte flora of both vegetations is an example, the ability to form both succulent tissues and salt-excreting epidermal bladder hairs being equally useful in both environments.

2 Wetland Soils: morphology and chemistry

2.1 Soil types

Soils are ascribed to *zonal* types which broadly correspond with geographical-climatic conditions but, throughout the world, poor drainage sometimes coupled with a high precipitation-evaporation (P/E) ratio, creates a range of *intrazonal* wet soils. The latter have a number of common characteristics related to oxygen deficiency and low redox potential which categorize them as *gley* soils. Irrespective of climate, gleys carry wetland vegetation types: marsh, swamp, fen and wet-heath world-wide, with swamp forests and cultivated rice paddy in the tropics (ETHERINGTON, 1982).

In extremely wet conditions organic matter accumulates faster than microbial respiration can dispose of it, creating organic *peat* soils or *histosols*, an *azonal* group which occurs anywhere and without any necessary relationship to surrounding zonal soil types. These peats are, most frequently, strongly acid mineral-poor bog soils with a specialized wet-heath vegetation dominated by peat-formers such as *Sphagnum* moss species, Cyperaceae and some grasses. Neutral or mildly alkaline peats are associated with *fen* vegetation which, at its richest is a species-diverse luxuriant herbaceous ecosystem which intergrades through a more acid, calcium-deficient *poor-fen* to acid peat bog vegetation. In late-succession, organic fen peats become colonized by a *carr* woodland of flood-tolerant trees and shrubs, charatateristically alders and willows (*Alnus* and *Salix* spp.) in N. America and Eurasia.

Peats are less frequent in the tropics because microbial metabolism is more rapid. They cover less than one percent of the tropical land surface and where they do occur are generally coastal or riverine groundwater peats with tidal mangrove swamp, reedswamp such as papyrus (*Cyperus papyrus*) or species-rich swamp forest. A few tropical peats are of the raised-bog type (see section 2.3), dependent on rainwater input, and clothed with a species-poor heath-forest.

Waterlogging may also be caused by tidal flooding, thus there is a group of gley soils and peats associated with salt marshes and tropical mangroves, their chemical characteristics being further complicated by salinity and the vegetation not only requiring tolerance of waterlogging but also halophytic characteristics.

2.2 Mineral soils – gleys

Soils are described by reference to their *profiles*, the sequence of layers (*horizons*) exposed in the wall of a soil-pit. Horizons form during soil

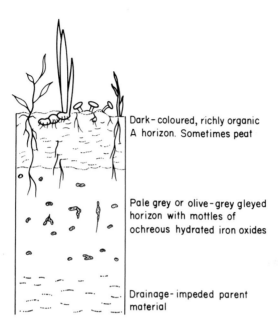

Dark-coloured, richly organic A horizon. Sometimes peat

Pale grey or olive-grey gleyed horizon with mottles of ochreous hydrated iron oxides

Drainage-impeded parent material

Fig. 2-1 A typical gley soil profile. The vegetation may be marsh or fen if the ground-water is reasonbably nutrient rich. Oligotrophic, acid ground-waters support wet-heath vegetation often with substantial peat formation.

development, are visibly different from each other, and have contrasting physical and chemical characteristics. The horizons of a gley soil are usually an upper organic rich or peaty layer overlying a paler coloured gley-horizon in which the background colour of grey or grey-green is frequently mottled with brown or ochreous patches of iron oxide enrichment (Fig. 2-1).

Gley soils have their pore space partly or completely flooded, for at least a portion of the year, and because oxygen diffuses some 10^4 times more slowly in water than in air, the dominant soil properties reflect shortage of oxygen. Organic matter fails to oxidize and accumulates at the soil surface while the deeper mineral soil is wholly or partly anoxic so that microbial oxidative metabolism has to depend on electron acceptors other than molecular oxygen (section 7.2). The most frequent of these are nitrate, Mn^{3+}, Mn^{4+}, Fe^{3+}, sulphate and carbon-dioxide. The characteristic grey-green colour of the gley horizon is due to the conversion of Fe^{3+} to Fe^{2+} by this process with loss of the normal red or brown oxide coloration. Fe^{2+} compounds are more water soluble than Fe^{3+} and may be washed-out by percolating water, thus bleaching the mineral soil (Fig. 2-1).

Few gleys remain permanently flooded and oxygen does enter during dry periods. It will also be shown later that living roots may leak oxygen to the surrounding soil. The result is that adjacent to roots, and in larger pores, Fe^{2+} re-oxidizes, precipitating visible mottles of iron oxide.

The chemical characteristics of the gley soil as a root environment, may be defined by the prevailing redox potential which will be a function of the ion species and organic compounds present and lies between 4–500 mV when oxygen is present and -150 mV when sulphate is being reduced to sulphide (Fig. 2-2). Once microbial respiration has scavenged all of the oxygen from the soil, its place as an electron acceptor is taken by NO_3^- which, in the process, is denitrified to N_2O or N_2. Waterlogged soils are thus nitrate deficient and their vegetation constrained to assimilate nitrogen as NH_4^+. Electron acceptance by the high valency states of iron and manganese then causes their concentration to increase in soil solution as the divalent compounds are more soluble in water than those of the oxidized forms.

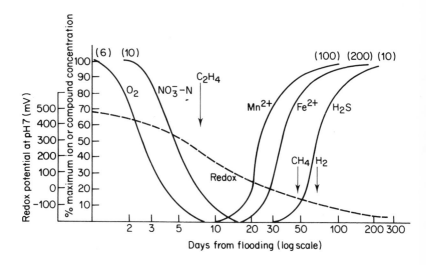

Fig. 2-2   A typical sequence of chemical changes following the flooding of a neutral, eutrophic soil. Except for redox potential (– – –)the curves represent changing concentrations of the named species in soil solution, plotted as a percentage of their maximum values which are indicated by the bracketed figures (mg 1^{-1}). The changes are consequences of microbial activity and comprise: respiratory oxygen reduction (O_2 + (CH_2O)$\to$ CO_2 + H_2O); nitrate respiration ($NO_3^- \to$ gaseous N_2 or N_2O); iron and manganese reduction (Mn^{4+} and $^{3+} \to Mn^{2+}$ and $Fe^{3+} \to Fe^{2+}$), the divalent species being much more soluble in water than the oxidized forms; and finally sulphate reduction ($SO_4^{2-} \to H_2S$). The vertical arrows indicate approximate thresholds for microbial evolution of ethylene, methane and hydrogen. Further discussion appears in section 7.2.

In organic-rich soils, and especially at high temperatures, sulphate may be used as an electron acceptor by sulphate-reducing bacteria, producing free hydrogen sulphide, In soils containing Fe^{2+} the sulphide is immediately removed from solution by the precipitation of very insoluble FeS which gives the characteristic black coloration to the *sapropel* sediments of eutrophic freshwaters and coastal mudflats: the presence of FeS is easily diagnosed by the odour of rotten eggs when the material is treated with dilute mineral acid.

Dissolved iron and manganese in many flooded soils reaches phytotoxic concentrations and, as described in Chapter 3, tolerance of these metals is a part of the niche differentiation of wetland plants. Sulphide is less commonly toxic, primarily because it is scavenged as FeS, though in soils which have been leached of Fe^{2+} by laterally percolating water, plant growth may be affected. The hot, organic-rich circumstances of rice culture are particularly conducive to sulphide formation and, though *Oryza sativa* is an extremely waterlogging-tolerant species, a range of its diseases are attributable to sulphide toxicity (Armstrong in ETHERINGTON, 1982).

When coastal salt marshes and mudflats are drained, or during the cultivation of rice paddy when the water level is lowered, FeS is oxidized by sulphur bacteria which liberate free sulphate (SO_4^{2-}). In calcareous soils gypsum crystals ($CaSO_4H_2O$) are formed and no pH change occurs, but in the absence of calcium the soil becomes extremely acid ($<$ pH 3) because sulphuric acid is produced. Large areas of acid sulphate soil or 'cat clays' have resulted from drainage of mangrove swamps in South-east Asia, South America and Africa (BRIDGES, 1978).

Anaerobic micro-organisms in waterlogged soil also liberate organic products which influence plant growth. When the soil first becomes anoxic some fungi produce ethylene (C_2H_4), a potent plant growth regulator. Organic acids such as acetic and butyric may also act as phytotoxins in anaerobic soils while oxidizable organic compounds may compete with plant respiratory oxygen need. Extremely reduced soils and sediments microbially generate copious amounts of methane (the inflammable 'marsh gas') and lesser amounts of hydrogen to the extent that the soil atmosphere above the watertable of some paddy soils may contain equal volumes of methane and nitrogen (section 7.2).

2.3 Organic soils

Peat-forming *mires* are most typical of high latitudes where low temperature limits both evaporation and microbial respiration but peat soils of all types occasionally extend even to the equatorial tropics when hydrological conditions are right. The circumstances of peat formation are three-fold (MOORE and BELLAMY, 1973) (Fig. 2-3). Primary peats are those which form in hollows receiving drainage water, and the accumulation of the peat actually reduces the water capacity of the basin. Secondary peats develop from primary by overspillage of the growing peat onto surfaces adjacent to the primary basin and because peat has a large water-storage capacity the total water retention of

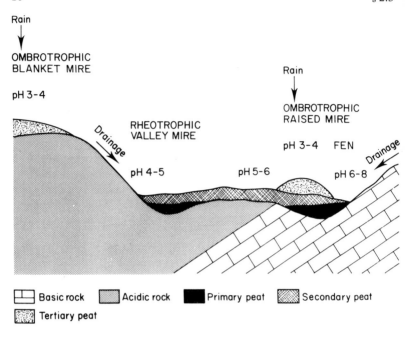

Fig. 2-3   The relationship of primary, secondary and tertiary peat formation to geology, hydrology and topography.

the surface is increased by its presence. Tertiary peats are independent of a pre-existing drainage basin and form on any level or gently sloping surface when precipitation substantially exceeds evaporation. Again, the water capacity of the surface is increased by storage in the peat.

Primary peats are acid and nutrient deficient if the catchment is oligotrophic but if the water supply is neutral or alkaline, fen-peat will form, thus, according to superficial geology, *valley mire* vegetation may range from wet-heath through poor-fen to rich fen (Fig. 2-3).

Secondary peats are similar to the primary peats with which they are associated as they remain in capillary contact with the underlying soil solution but, with a sufficiently high precipitation evaporation ratio, tertiary peat growth involves loss of hydrological contact with underlying system. Peatlands developed in this way over extensive land surfaces are described as *blanket mires*. More locally, the continued growth of peat over a valley mire creates a tertiary *raised mire*, typically of a domed cross-section because resistance to lateral water seepage results in a convex watertable which enhances peat growth near the centre of the system.

Blanket mires which are independent of drainage-hydrology are the most widespread of peatlands, covering more than a million square kilometres of the northern hemisphere; over 1% of the land-surface. Together with the tertiary peats of raised mires these blanket systems rely on solutes in precipitation and blown dust for their nutrition and, in consequence, are the world's most oligotrophic terrestrial surfaces. For this reason they are described as *ombrotrophic* (Gk. *ombros*: a rainstorm).

The primary and secondary peats of valley mires are less nutrient deficient, remaining in capillary contact with flowing groundwater. These are the *rheotrophic* peats (Gk. *rheos*: stream) and their peat growth is a function of wetness and anoxia alone without the further effects of nutrient deficiency and acidity. Alkaline fen-peats do in fact oxidize very fast when drained, hence the great wastage or land surface-level during the past two centuries in the English fenlands (GODWIN, 1978).

Peatlands are less common in the tropics but rheotrophic mires do occur on riverine flood-plains and coastal mangrove swamps. If precipitation greatly exceeds evaporation, ombrotrophic peats may form. Raised bog occurs in the East Indies vegetated with a species-poor swamp forest resembling the heath-forests of tropical podzols and sometimes almost monospecifically dominated

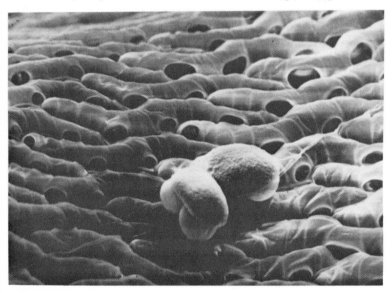

Fig. 2-4 A scanning electron micrograph of a *Sphagnum* moss leaf surface with two trapped pine pollen grains (*Pinus sylvestris*). Each grain is about 60 μm in length. Note the large open pores of the empty hyaline cells which act as a capillary sponge, retaining water as the bog grows by accumulation of plant remains to form peat. The pollen grains are preserved as the vegetation forms peat, their characteristic shapes and surfaces sculpturing permit identification and give clues to the prevailing vegetation at the time of preservation. (Preparation by Mrs. V. Rose.)

by the giant buttress-based dipterocarp tree, *Shorea albida*. The porridge-like peat, 7 m or more in depth, is mainly derived from wood and tree litter. It is very acid (*c*. pH 3) and feeds oligotrophic 'blackwater' streams akin to those of heath forest. The underlying mineral soil is often a bleached, iron-depleted clay closely resembling the 'seat earths' of coal seams which of course originated from tropical swamp forest.

The early stages of ombrotrophic peat formation at high latitudes depends on the ability of the mire to maintain its own perched watertable even when it overlies a permeable substratum. Mosses of the genus *Sphagnum* predominate in this process because their microporous structure gives them great capillary water-holding power. Each leaf, and the stem cortex, contains a large proportion of dead *hyaline* cells the surface of which is perforated by pores (Fig. 2-4). Even in drought conditions water may be squeezed from tussocks of apparently quite dry *Sphagnum*. In addition to this water-holding ability the leaves of *Sphagnum* are highly efficient cation exchangers, scavenging metals from percolating water which is acidified by the exchange of hydrogen ions for its dissolved metal cations. *Sphagnum* is a high latitude genus but does occur in tropical montane mires and is recorded from the transition between reedswamp and the wet-soil palm, *Phoenix reclinata*, in Ugandan swamp lands.

3 Ecophysiology of Wetland Vegetation

3.1 The wetland problem

The chemical characteristics of waterlogged soils are harmful to dryland species but the niche is opened to a wide range of wetland plants by adaptive modifications. Soil oxygen deficiency demands either a transport system which can supply it to underground organs from aerial structures or anaerobic metabolism must prevail while toxic concentrations of Mn^{2+}, Fe^{2+}, S^{2-}, H_2S and organic compounds require either a modified metabolic chemistry which continues to function in this hostile environment or the plant must be armed with detoxifying mechanisms. Wetland plants must also differ from their dryland cousins in absorbing nitrogen as the NH_4^+ cation because dentrifying micro-organisms scavenge nitrate from anaerobic coils.

Plants inhabit soils which range from well-drained to permanently flooded (Fig. 3-1). In freely draining soil (a), waterlogging causes no problems though even here the soil pores may be totally water-filled for many hours after rain, causing slight anaerobiosis in the centres of soil crumbs. This may occasionally be beneficial as low redox potential increases the solubility of iron, and perhaps relieves its deficiency in alkaline soils.

Plants with some roots in a waterlogged soil (b) may have to cope with toxicity but water and nutrient uptake will be normal in those roots above the watertable. If the water level coincides with the soil surface (c), the problem of toxicity is compounded by lack of an external oxygen supply and once the water level rises above the soil (d) and particularly when the shoot is fully submerged (e) the misfortune is exacerbated by the low solubility of oxygen in water and its much slower diffusion in the liquid phase.

Plants of such soils may be stress-avoiders, either having their lateral roots very near better-oxygenated soil surfaces or developing superficial adventitious roots in response to flooding. Alternatively they may tolerate the stress by having morphological and anatomical features which allow internal oxygen transport to satisfy the needs of root respiration and also the oxidation of chemically-reduced toxins. Such characteristics must also be linked to biochemical adaptations which give tolerance of the inevitable periods of total anoxia during flooding.

3.2 Stress avoidance: dormancy or surface rooting

In high latitudes the wet season usually coincides with winter when adult plants are dormant or most annuals exist only as seeds. Such dormancy minimizes the problem of soil oxygenation while the deciduous habit limits not

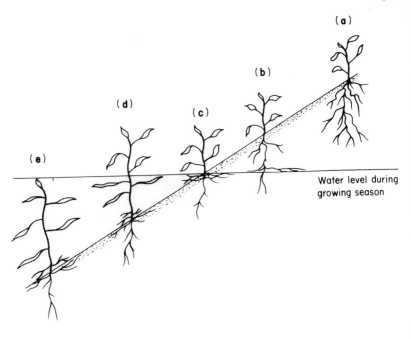

Fig. 3-1 The relationship between plants, soil and watertable ranging from the dryland environment (**a**) through varying degrees of waterlogging of the roots (**b** and **c**) to partial or total flooding of the shoot system (**d** and **e**).

only water uptake but also ingress of reduced toxins. This relationship of dormancy to waterlogging appears, however, to be spurious correlation as the waterlogged soil is never so hostile as to *impose* dormancy: reedswamp plants grow freely in microbiologically active, strongly reducing soil and some tropical forest species remain in full leaf when flooded to depths of a metre or more.

Production of adventitious 'water roots' close to a floodwater surface is common in many waterlogging-tolerant species. A remarkable photograph is reproduced in TANSLEY (1939, Plate 12) showing a large tussock of such roots induced, over one metre above ground, by natural flooding of *Salix alba* (white willow). In woody plants such as the willows and the N. American blackgum (*Nyssa sylvatica*) the flooding first causes hypertrophy of lenticel tissue to form a callus from which adventitious roots then emerge close to the better oxygenated air–water interface.

Many herbaceous plants behave similarly, for example, the flood-tolerant hairy willow herb (*Epilobium hirsutum*) has permanently visible root initials for several centimetres above the stem base but its intolerant congener, the rosebay willow herb (*Epilobium angustifolium*) has no such initials and fails to produce adventitious surface roots on flooding (Fig. 3-2). It has been shown

Fig. 3-2 Flood-induced adventitious roots on the stems of great hairy willow-herb (*Epilobium hirsutum*). The narrow-leaved willow-herb (*Epilobium augustifolium*) is intolerant of waterlogging and is unable to form adventitious roots (left hand side of photograph). Both plants were flooded for 5 days to the level of the leaf whorl shown. The epinasty and curling of the leaves is typical of ethylene (ethene) injury (section 3.5).

experimentally that formation of such roots permits more rapid recovery from waterlogging injury.

Near-surface lateral roots may be associated with other morphological features, for example the 'knees' or pneumatophores of some mangroves (e.g. *Avicennia germinans*) and of the swamp cypress (*Taxodium distichum*). These are very porous vertical outgrowths of lateral roots which project for several centimetres above the soil surface where they are furnished with lenticels. Each pneumatophore carries a cluster of thin absorptive roots just below the soil surface and, in the case of the mangrove *Avicennia germinans*, new layers of surface root grow as sediment accumulates. These mangrove pneumatophores are shown in Fig. 1-2(a) together with the stilt roots of the black mangle (*Rhizophora mangle*) which are also furnished with shallow absorptive roots and above-water lenticels.

3.3 Stress tolerance: morphology, anatomy and oxygen diffusion

The first plant anatomists noted that aquatic plants were unusually porous, their tissues containing enormous intercellular spaces often forming regularly arranged chambers. Physiologists now refer to such tissues as aerenchyma (Fig.

3-3) though this term was originally coined to describe a rather specialized porous tissue arising secondarily from a phellogen in the epidermal or cortical layer.

Roots in waterlogged soils are similarly porous and the air-space formation is environmentally inducible, thus rice roots grown in waterlogged soil contained 45% air space, almost twice as much as those in freely-drained soil (Armstrong in ETHERINGTON, 1982). Even dryland species, which are inherently much less

Fig. 3-3 Chambers of aerenchyma is a longitudinal section of bullrush stem (*Schoenoplectus tabernaemontani*). Stem diameter is 12.5 mm. Collected from reedswamp near South Wales coast.

porous, show such responses, for example barley (*Hordeum vulgare*) roots produce less intercellular space in aerated than non-aerated nutrient solution.

Early workers simply assumed that the role of pore space was to supply oxygen to buried organs embedded in 'asphyxiating mud' but the continuity of the air space and the dependence of root upon shoot for oxygen supply was not convincingly demonstrated until about 40 years ago (ARMSTRONG, 1980).

The use of tracers such as $^{15}O_2$ has shown, even in dryland plants, that oxygen can diffuse to the root system. This oxygen not only supports root respiration but may also leak to the exterior in anaerobic conditions, the radial oxygen loss being demonstrable with redox dyes such as reduced indigo-carmine which becomes blue on contact with oxygen (Fig. 3-4) or dissolved

Fig. 3-4 Radial oxygen loss from roots of willow-herb (*Epilobium hirstum*) made visible by oxidation of reduced indigo carmine. The grey haloes around the roots show the oxidation of the pale yellow reduced dye to its bright blue oxidized form. The dye solution is diffusion-stabilized with a 1% agar gel which is used to flood the root system before it sets (about 30ºC). The rate of radial oxygen loss may be calculated by measuring the extension rate of the oxidized areas. This photograph was taken 10 minutes after commencement of oxidation.

oxygen may be measured by a suitable electrometric technique (Armstrong in ETHERINGTON, 1982).

If roots are excavated from waterlogged soils they are often stained brown by Fe^{3+} hydrated oxides which may even form hard tubular coats with walls several millimetres thick. These deposits form by oxidation of Fe^{2+} in solution and precipitation of the less soluble Fe^{3+} hydroxy-oxides together with manganese compounds which may behave in the same way.

The significance of these observations is that Fe^{2+}, Mn^{2+} and other reduced materials are phytotoxic and their precipitation, when they meet oxygen leaking from intercellular space, serves as a protective mechanism. Electron probe micro-analysis shows that Fe^{2+} was oxidized not only on the surface of rice (*Oryza sativa*) roots but also in the large cortical aerenchyma spaces thus limiting uptake into the vascular system. The wide, strongly porous cortex of tolerant species may be interpreted as an oxidizing 'filter' which ensures maximal exposure of the transpiration stream to intercellular oxygen.

The apparent xeromorphy of some wetland plants, particularly those of oligotrophic soils, was originally interpreted as a response to 'physiological drought'. This concept has long been abandoned but there is now some experimental support for the suggestion that a reduced transpiration rate will slow the movement of reduced toxins to the root surface and permit more effective oxidative detoxification of Fe^{2+}.

Plants of the reedswamp environment such as the widespread reed-grass (*Phragmites australis*), cattails or reedmaces (*Typha* spp.) rushes (*Scirpus* spp.), complete their growth in the most severe of all waterlogged environments, their roots and rhizomes in anaerobic mud and their leaves in a very exposed, often sunlit environment. In such plants the transpiration stream must transport substantial amounts of reduced toxins to the root surface. It has been suggested that the monocotyledon habit of growth is particularly suited to this situation: their free adventitious rooting and frequent absence of stem tissue between roots and leaf easing the oxygen transport problem. It may be this facility of the monocotyledons which has created the great visual similarity between reedswamps from the tropics to the tundra.

An alternative interpretation of aerenchyma is that of a mechanical *cum* metabolic adaptation in which the chambered structure provides a mechanically strong tissue using a minimum of respiring tissue. This would be advantageous in an oxygen deficient environment and might explain the enigmatic overdevelopment of aerenchyma in some organs which is difficult to explain in terms of oxygen diffusion alone as transverse septa must often be the limiting factor.

3.4 Tolerance of toxins in anoxic soil

Despite the defence mechanism of radial oxygen loss, wetland plants must sometimes encounter periods when the system fails if the whole plant is flooded. Furthermore, the apical tissues of roots are unvacuolated and without pore space so that any oxidative protection must come from some distance behind the root tip. It is likely then that plants of waterlogged soils will have acquired adaptive tolerance of reduced inorganic and organic toxins which cannot always be excluded.

The two heathers, *Erica cinerea* and *E. tetralix*, respond very differently to waterlogging, *E. cinerea* being killed within a few weeks and *E. tetralix* surviving indefinitely (JONES and ETHERINGTON, 1970). Tissue analysis showed that both species immobilize a large amount of iron in the roots and though *E. tetralix* was more efficient in intercepting iron its leaf tissue still became considerably iron-enriched following waterlogging.

To test the hypothesis that *E. tetralix* is inherently more tolerant of iron the authors supplied cut shoots of the two species with Fe^{2+} sulphate at 1, 10 and 100, μg ml^{-1}. Control shoots in water remained healthy for 27 days but *E. cinerea*, at any concentration of Fe^{2+}, died before the end of the experiment. *E. tetralix* was uninjured, even by 100 μg ml^{-1} Fe^{2+} though some of the 10 μg ml^{-1} shoots looked unhealthy after 27 days. These results suggest that tolerance of iron at the cellular level may play its part in differential response to waterlogging, and other workers, using solution culture techniques, have confirmed these findings for both iron and manganese in circumstances where oxidative exclusion from the root is unlikely.

The sensitivity of shoot tips to Fe^{2+} may be investigated by incubating them in suitable anaerobic solution culture and, as might be expected, is related inversely to waterlogging tolerance (N. Smirnoff, 1981, pers. comm.). This technique has also been used to compare populations of *Carex flacca* (carnation sedge) from different habitats: those from wet sites were less sensitive to Fe^{2+} toxicity than dry site populations (M.S. Davies and C. Kunz, pers. comm.).

3.5 Metabolic and biochemical responses to flooding

Wetland plants in extreme conditions may have to cope with long periods of total anoxia, thus floating-leaved and reedswamp species which lose their leaves during winter must resume growth and produce considerable new tissue before reaching a copious oxygen supply. They are forced to rely upon anaerobic metabolism and it was in such plants that the accumulation of ethanol was first noted. The freshly dug rhizomes of the waterlily (*Nuphar advenurn*) not only have an aromatic-esteric odour but yield sufficient ethanol on distillation to be identified by fairly crude analytical techniques (LAING, 1940). The rhizomes also produced ethanol whenever the external oxygen concentration was less than 3% v/v and the same is true of the leaves of both this species and the reedswamp cattail or reedmace (*Typha latifolia*) which are frequently flooded after rain.

Laing suggested that tolerance of anoxia might be given by the ability to ferment carbohydrate and it is now known that many plants convert glycolytically-produced pyruvate to ethanol *via* acetaldehyde. To accumulate any significant oxygen debt substantial amounts of ethanol must be formed: this is certainly the case in some species; for example, yellow water iris (*Iris pseudacorus*) rhizomes transform almost 80% of their respired carbohydrate to ethanol during short periods of anoxia and of the remaining 20% a detectable amount was converted to shikimik acid (usually a precursor of the aromatic amino acids).

The accumulation of shikimik acid and end-products of glycolysis other than ethanol has been interpreted as an adaptive waterlogging-tolerance mechanism (CRAWFORD, 1976), ethanol being presumed to be toxic at high concentration. It is likely that a further adaptation to survival of waterlogging will be either tolerance of ethanol or provision for its loss by leakage in solution and through the intercellular spaces as vapour.

Crawford's research school has proposed that waterlogging-tolerant plants control ethanol toxicity by limiting the rate of glycolysis, by diversification of its end-products as shown in Fig. 3-5 and by provision of carbohydrate reserves to cope with the low energy yield of anaerobic metabolism. This has, however, been challenged as some workers have failed to show accumulation of end-products other than ethanol in tolerant and intolerant species (e.g. SMITH and AP REES, 1979) or have shown malate to accumulate independently of ethanol, as did KEELEY (1979) for flood tolerant blackgum (*Nyssa sylvatica*). Keeley suggested an alternative role of malate as an adjuster of cellular ionic balance in

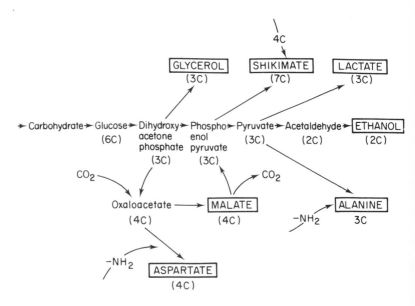

Fig. 3-5 A scheme for the diversification of the end-product of glycolysis. Compounds in boxes have been recorded as accumulating in plants under anoxia. Numbers in brackets indicate carbon atoms per molecule. (After CRAWFORD, 1976.)

a situation where CO_2, Fe^{2+} and Mn^{2+}, at least, are above normal concentrations. Such suggestions deserve further research because malate may have a role as a chelating agent in heavy-metal tolerance and could play a part in protecting waterlogged plants from Fe^{2+} and Mn^{2+}.

The growth-controlling metabolism of plants may be changed by flooding. Normal plant tissues synthesize a small amount of ethylene (ethene; C_2H_4) from methionine *via* intermediates in a cyclic process. Low concentrations of ethylene interact with other growth-controlling substances as a necessary part of normal metabolism. Surplus ethylene escapes through the intercellular space and stomata but flooding effectively seals-in the gas and, at the same time stimulates its production (DREW *et al.*, 1979). The leaf curling and epinasty (petiole drooping) shown in and often described as a first symptom of flooding injury, is produced either by this effect or by exogenous microbial production of ethylene in the anoxic soil. High ethylene concentrations initiate adventitious root production and also the lysigenous formation of intercellular space, both interpretable as adaptive responses to waterlogging (DREW *et al.*, 1979).

Flooding inhibits root growth in plants which have an inherently high ethylene production but flood-tolerant species such as rice (*Oryza sativa*) produce less and are not so seriously affected. Shoot extension is usually inhibited by ethylene but in some floating-leaved aquatic plants it may be stimulated thus restoring leaf contact with the aerial environment; a useful sensing mechanism and adaptive response.

Total flooding, involving the shoot system, may initiate a further sensing mechanism because the ratio of red to far-red light (660 nm v 730 nm) is greatly increased by passage through water (3 to 4 times for a one metre water column). Evidence now exists that this controls heterophylly in *Hippuris vulgaris* (mare's tail) which produces very thin leaves with almost no cuticle below water and mesophytic leaves above. This is almost certainly a phytochrome-mediated response and may well play a part in other flooding responses such as petiole elongation in floating-leaved aquatics.

3.6 Conclusion

A plant rooted in a soil with an oscillating water-table cannot have an optimum 'strategy' as the adaptive mechanisms giving waterlogging tolerance either cannot respond fast enough or, if they are already present, have a significant 'cost' while the soil is well drained. This cost may be in the greater drought sensitivity of plants with very porous roots (KEELEY, 1979) or possibly the thicker, softer aerenchymatous roots may be damaged in penetrating a well-drained soil which is mechanically much stronger than its flooded counterpart. In KEELEY'S (1979) words: 'perhaps the only optimum strategy in these circumstances is one of compromise'. The chemical and physical consequences of waterlogging are so numerous and the responses of plants so varied that a host of adaptations have evolved. Some attempt to present these in simplified form is made in Fig. 3-6.

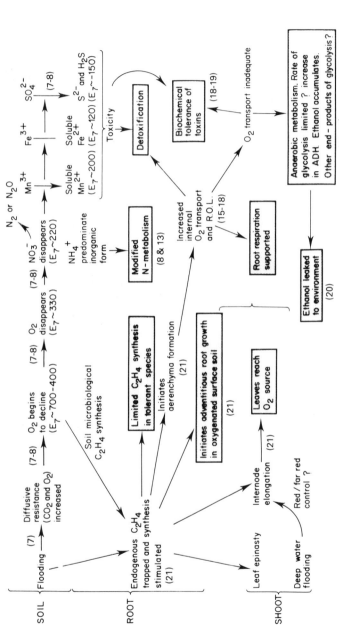

Fig. 3-6 A simplified scheme for waterlogging tolerance. Arrows indicate sequences of events and boxes represent adaptive consequences. E_7 - redox potential corrected to pH 7.0 (mV). R.O.L. - radial oxygen loss. A.D.H. - alcohol dehydrogenase. Question marks indicate mechanisms which have not been fully investigated or about which there is controversy. Bracketed numbers are page references.

4 Animals in the Wetland

4.1 Invertebrates

Small invertebrates inhabit the water film between soil particles and for this reason there is an overlap of species composition between open water and soil. Many Protozoa, Rotiferae and Nematoda fall into this category, occurring in both organic-rich water and in wet-soil litter. Similar overlaps are found in the larger invertebrates: the turbellarian flatworm *Prorhynchus stagnalis* lives in water, wet mats of blue-green algae (Cyanobacteria) and in wet soil while aquatic detrivores such as the freshwater shrimp (*Gammarus pulex*) and the water louse (*Asellus aquaticus*) both abound in beds of leaf litter on wet soil but almost always adjacent to open water.

To some extent these are exceptions: most of the soil fauna are susceptible to loss of air-filled pore space, anoxia of soil water or the chemical toxicity of the reducing environment in waterlogged soil. The ecology of soil invertebrates is quite strongly dependent on these differences between wet and well-drained soils. The upper vegetation canopy, grass tussocks and heaps of litter above flood level are more like a dry terrestrial habitat and its inhabitants are less strongly differentiated from those of adjacent ecosystems. Small differences of level are reflected in large variations in soil wetness, plant species and soil organic matter thus permitting a very species-rich cohabitation of hygrophilous and xerophilous organisms.

The invertebrates of the wetlands range from obligate to facultative forms which are present for the following reasons:
1) Part or all of the life-cycle is aquatic.
2) The organism requires high humidity or free water.
3) A specific wetland food plant or animal is required.
4) A specific habitat physiognomy is required: this may or may not overlap with dryland ecosystems.
5) The wetland provides a refuge from predation or competition in the absence of which the organism might occur elsewhere.

The microfauna of many well-drained soils is dominated by Collembola (spring-tails) which are apterygote insects and by Acari (mites) belonging to the Arachnida (spiders). The density of these organisms often exceeds a tenth of a million individuals m^{-2}. Mesophile and xerophile forms breath by means of an open tracheal system and consequently suffer damage by flooding unless sufficient air is trapped in body surface hairs to form a 'physical gill'. The more hygrophilous species, however, have a much thinner cuticle, are unable to control water loss in dry conditions and respire by diffusive gas exchange through the whole body surface. Examples of wetland species are the springtail *Isotomurus palustris*, abundant in waterside soils and the cryptostigmatid mite, *Hypochthonius refus* which is found in *Sphagnum* moss. In general both numbers and species diversity of Collembola and of prostigmatid mites declines with wetness while cryptostigmatid mites tend to increase in relative abundance.

Some invertebrates have solved the problems of oxygen availability, not by carrying a surface film of air trapped in body hair, but by maintaining an air-filled burrow. This is well developed amongst fiddler crabs (*Uca* spp.) of salt marshes and mangroves which may pass the high tide period in such a burrow to which they also rush when frightened.

Other sources of oxygen are available in anaerobic soil and sediment for example larvae and pupae of two very different insects, mosquitoes such as *Mansonia* spp. and chrysomelid beetles of the genus *Donacia*, both tap the aerenchyma of reedswamp plant roots using a hollow spine-like siphon (Fig.

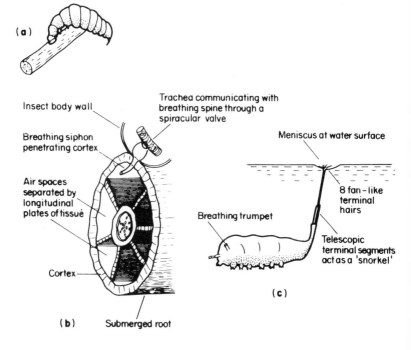

Fig. 4-1   Air-breathing adaptations of swamp-dwelling insects. (**a** and **b**) Beetle larvae of the genus *Donacia* tap the root intercellular space of reed-swamp plants such as *Typha* spp. (reedmace or cat-tail). (**c**) The 'rat-tailed maggot' larvae of syrphid flies belonging to the genus *Eristalis* lies in anaerobic liquid mud, its telescopic breathing siphon breaking the surface through a fan of hydrophobic hairs.

4-1). Telescopic siphons which may be extended to reach the surface of liquid mud or anoxic water have evolved separately in the larvae of drone flies (*Eristalis* spp.) and in crane flies of the genus *Ptychoptera* (Fig. 4-1).

Several unrelated organisms have become adapted to low partial pressure of oxygen by increased affinity, using the haemoglobin pigment more familiar as a component of mammalian blood. Such diverse groups as tubificid and lumbricid Oligochaete worms, some pulmonate molluscs (*Planorbis* spp. and Insect midge larvae of the genus *Chironomus* have separately evolved this mechanism.

The haemoglobin content of lumbricid earthworms is correlated with habitat wetness, thus the surface-dwelling *Lumbricus rubellus* has 14 mg g^{-1} body weight compared with 39 mg g^{-1} in *Eiseniella tetraeda* which lives in flooded soil. Intermediates between these extremes have correspondingly intermediate haemoglobin values.

Tubifex worms inhabit mud tubes in sapropel sediments so reducing that they contain several percent of iron sulphide (FeS), the tails extending, as a pseudo-gill, into the more oxygenated water above. The amount of tail protruded is negatively correlated with the oxygen content of the water. These are the worms sold by aquarists in knotted masses as fish-food. *Chironomus* midge larvae inhabit the same type of sediment together with rams-horn snails (*Planorbis* spp.), coloured bright red by haemoglobin, as an association of organisms which is very characteristic of organic-polluted waters of high biological oxygen demand.

The Oligochaete worms of freshwater environments are replaced by Polychaetes in coastal habitats and in some situations these organisms have as much impact on soil-formation as lumbricid worms in terrestrial ecosystems. *Arenicola marina* (lugworm) maintains a ventilating water current through its burrow and when this is in reducing sediments the oxidation of Fe^{2+} causes the walls of the burrow to be stabilized with a precipitated concretion of ferric oxide several millimetres thick. These hard tubular structures are a characteristic of the coastal soil type *sandwatt gyttja*.

Relatively little is known of animal responses to the other reduced chemicals which accumulate in waterlogged soil though KUHNELT (1961) noted that many soil animals are able to tolerate 0.5% hydrogen sulphide for seven days, suggesting that they are much less sensitive than roots. A few animals which are intolerant of soil wetness (e.g. *Porcellio scaber*, woodlouse) are killed by such treatment. Some animals also respond differentially to carbon dioxide, for example deep-dwelling Collembola are less sensitive than surface-dwelling species.

In the last chapter it was suggested that roots might survive anoxia by anaerobic metabolism and accumulation of an oxygen debt. Three such mechanisms were first described in various intertidal bivalve molluscs and annelids. Firstly, organisms which endure longish periods of anoxia show a metabolic regulation which includes a general reduction of rate and also avoidance of a Pasteur effect (the increased pace of glycolysis, at low oxygen concentration, shown by normal aerobic organisms). Secondly the products of glycolysis are diversified, presumably avoiding the toxic accumulation of a single compound such as ethanol (see Fig. 3-5). Finally, because anaerobic metabolism is energy-profligate, these tolerant organisms have additional stores of respirable carbohydrate; usually glycogen in animals.

Little is known of these mechanisms amongst waterlogged soil invertebrates and they might well be a rewarding subject of investigation.

Many wetland gastropods are of aquatic origin, for example the marsh snail (*Lymnaea glabra*) and the dwarf marsh snail (*L. truncatula*) both wander freely from freshwater into adjacent marsh grassland. The latter snail is the second host of the sheep liver-fluke (*Fasciola hepatica*) and its wandering habit makes many marshy pastures unsafe for sheep. Periodically-flooded litter in fens and reedbeds has other characteristic non-aquatic snails, for example *Succinea putris* (amber snail) and it also differs in the relative numbers of other invertebrates. Dipterous larvae are particularly common in wet litter and various isopods are adapted to this habitat, for example the British woodlouse *Ligidium hypnorum* which typically dwells in reedswamp and fen tussocks and is not modified for air-breathing as are its more terrestrial relatives which are furnished with pseudotrachea in the abdominal limbs (pleopods).

Wetlands support a variety of bloodsucking insects of which the larvae inhabit water, wet moss and litter. The most familiar are, perhaps, the mosquitoes (Culicidae) which lay eggs on water, or in the case of some *Aedes* spp., on wet soil in depressions which become rain-filled. A few species, for example, *A. detritus* and *A. caspius* are specifically associated with saltmarsh pools. The larger biting horse-flies (tabanids) also have larvae which inhabit water and wet moss.

Other carnivorous insects occur at the land–water margin, for example the shallow water of reedswamps is a half-submerged jungle in which the tigers are the nymphs of dragonflies (Odonata), some able to consume creatures larger than themselves, even small fish. The mature nymphs emerge from the water by climbing a reed stem and, within two or three hours, the final moult has produced the adult dragonfly or 'mosquito hawk'. The loudly rustling wings and purposive quartering of territory by the larger anisopterid dragonflies together with a myriad hovering metallic blue, green and red thread-bodies of the zygopterid damselflies are perhaps more evocative of the waters-edge than any other sight or sound.

The canopy of wetland vegetation provides conditions not much different from that of any other plant cover, thus the hunting spider (*Lycosa nigriceps*) is equally at home in deep mixed-fen, in gorse-heather heath or in sand-dune marram grass. By contrast, some canopy dwellers are surprisingly specific to wetlands: the North European marsh grasshopper (*Stethophyma grossum*) always occurs on peaty substrata despite the fact that it feeds on the canopy well above flood level.

This survey must inevitably be incomplete, so great is its scope: nothing has been said of the Coleoptera (beetles) some of which are aquatic throughout their lives, some have only aquatic larvae, some are associated with specific wetland plants and yet others universally use vegetation cover as their hunting ground. Similarly the Lepidoptera (butterflies and moths) have been omitted despite the ecological and conservational interest of species such as the extinct swallowtail (*Papilio machaon*), in Britain, which are threatened by foodplant loss as wetlands are increasingly drained for agriculture and industry (see section 9.4). More information may be found in ELLIS (1965), KUHNELT (1961), JACKSON and RAW (1966), ANDREWS (1973), WALLWORK (1976) and DYKYJOVA and KVĚT (1978).

4.2 Vertebrates

Many vertebrates invade wetlands from the surrounding terrestrial habitats, thus in Europe the grass snake (*Natrix natrix*) is a familiar hunter of frogs and insects in the reedswamp margin and is also a capable swimmer and yet it is equally at home pursuing the small mammals and other denizens of dry grassland. Many of these grassland mammals also penetrate the upper reedswamp: amongst the herbivores, the harvest mouse, common vole and bank vole (*Micromys minutus, Microtus arvalis* and *Clethrionomys glareolus*) are common while the insect-hunting shrews (*Sorex* spp.) differentiate little between reedswamp and tall grassland.

Wetter reedswamp with open water pools provides the habitat of other more nearly obligate wetland vertebrates of which the most familiar are the frogs and toads (e.g. *Rana* and *Bufo* spp.), amphibia which depend on water as a habitat for their tadpoles but otherwise roam and hunt throughout the wetland and its terrestrial fringes also comprising a large part of some tropical forest-floor faunas. Many reptiles including numerous species of water-snakes (sub-family Homalopsinae) and the sub-aquatic turtles (super-family Testudinoidea) are charactristic of the water-edge throughout the world.

A few species of fish are adapted to the mud and shallow water-swamp environment, coping with water so deoxygenated that it would asphyxiate normal fish. Most typically they belong to the fauna of seasonal tropical swamps which occassionally dry out completely and may impose enormous fish mortality (Bonetto in HASLER, 1975). Those which survive may aestivate or migrate overland and belong to the polygeneric group of air-breathing fish. Six independent means of air-breathing amongst eight species from the Paraguayan Chaco have been found and many other gill-respiring species draw surface water over their gills and drown if confined to deeper water.

The swamp-dwelling catfish (*Clarias mossambicus*) is not only an air-breather capable of migrating overland but also possesses a blood haemoglobin which has an exceptionally high oxygen affinity combined with tolerance of large concentrations of carbon dioxide in the blood. The lungfish (*Protopterus aethiopicus*), another African swamp species, similarly tolerates high blood carbon dioxide, air-breathes through its modified swim-bladder and aestivates during drought in a mucous cocoon. A few temperate zone fish are also able to migrate overland, for example the eel (*Auguilla anguilla*) which shows some respiratory gas exchange through its skin.

Amongst the mammals the water vole (*Arvicola amphibius*) of northern Europe and the North American muskrat (*Ondatra zibethicus*), sometimes feral in Europe, are both herbivores, often heavily grazing *Phragmites, Typha* and *Glyceria maxima* reedswamp. In North America, habitat destruction by muskrat 'eat-out' of reedswamp creates flooded, mud-churned 'creveys'. Similar eat-outs by geese (*Anser* spp. and *Branta* spp.), in North American saltmarsh, cause pool formation; damage sometimes compounded by muskrats.

The marvellous adaptation to amphibious life, well seen in the gentle cruising of water voles amongst the floating duckweeds of reedswamp pools is found in other wetland mammals, for example the beavers (*Castor* spp.). These rodents are adept swimmers and create swamp pools by felling small trees to build dams which house their 'lodges'. So effective is this habit that beavers are suspected of having caused widespread swamp formation in North America and in Europe prior to the advent of man.

Despite the remarkable adaptations of these creatures temperate-zone reedswamp remains more or less inaccessible to many larger predatory animals though the mink (*Lutreola lutreola*) and otter (*Lutra lutra*) seems as much at home in water as in the dried parts of reedswamp. Despite these occasional hunters, the reedswamp provides cover and refuge particularly for nesting water birds which either construct nests among the reed stems (reed-warblers; *Acrocephalus* spp.) or on a mound of felled reeds, hidden from view and shaded by the surrounding vegetation (swans and ducks; *Cygnus* spp., *Anas* and other genera). Many of these reedswamp-nesting and roosting birds hunt over the adjacent waters or tidelands, thus the reed-warblers feed largely on flies whose larvae are detrivores in the adjacent water. Saltmarshes often provide cover and nesting ground for birds which feed on the adjacent tidal flats, often relying on huge populations of a few species such as the algal-feeding spire shell (*Hydrobia ulvae*) in Europe. In some situations these birds may close the circuit of nutrient cycles between the coastal waters and the shore: mangrove litter exported to estuarine waters is eaten by *Littorina scabra* (periwinkle), which in turn feeds mangrove-roosting waders and returns exported nutrients to the producer community.

Tropical reedswamps support much larger herbivorous mammals: in Africa the hippopotamus (*H. amphibius*) feeds voraciously on the grass *Vossia cuspida* in company with an antelope, the puku (*Kobus vardoni*), sometimes causing habitat damage by trampling and wallowing. The rather long, high-rainfall wet season imposes an interesting cycle of a animal migration in these riverine and lakeside tropical swamp grasslands. Another African antelope, the lechwe (*Kobus leche*), also feeds in permanently flooded areas while other large mammals which cause bank and marginal vegetation damage are the South American capybara (*Hydrochaerius capybara*) and the tapir (*Tapirus terrestris*). The Amazonian manatee (*Manatus inunguis*) feeds so voraciously on aquatic vegetation that it has been suggested as a biological control for water hyacinth (*Eichhornia crassipes*) infestation. The reedswamp and forests of the warmer parts of the world also support large carnivores of which the aquatic reptiles; alligators and crocodiles (*Alligator, Crocodilus* and other genera) are the most familiar. These reptiles essentially prey on animals which are attracted to the water margin to drink, fish or feed. Amongst the reptiles a few aquatic snakes are also represented.

5 Mineral Nutrition of Wetland Vegetation

5.1 Rainfed and groundwater systems: contrasting environments

Nutrient availability in wetlands ranges very widely: rainfed peats are probably the world's most mineral deficient biotopes while some groundwater fens and reedswamps receive continuously replenished mineral supply and also maintain conditions which are very favourable for nitrogen fixation (Fig. 5-1).

The vegetation of oligotrophic peat is superbly adapted to mineral deficiency having efficient scavenging systems for trace elements, a low requirement for the macronutrients and conservational mechanisms such as evergreen leaves which limit loss of nutrients. Many peat soils are so deficient, even of trace elements, that problems arise when they are used for agriculture or forestry. A typical example is the 'reclamation disease' caused by copper deficiency of North American peats after liming and fertilization.

Some eutrophic wetland species, particularly those of reedswamp and shallow water, are capable of enormous growth rates and, in favourable climates, they are amongst the most biologically productive of ecosystems (section 6.2). Such prolific growth frequently causes blockage of waterways, drainage and irrigation channels if they recieve extra nitrogen and phosphorus from fertilizer leaching and domestic sewage. This same response has also been put to use in the tertiary treatment of sewage to remove the offending nitrate and phosphate. In one pilot scheme *Scirpus palustris* (bullrush) in an artifical marsh was able to retain up to 20 g P m^{-2} y^{-1} (Sloey *et al*. in GOOD, WHIGHAM and SIMPSON, 1978).

5.2 Nitrogen fixation and denitrification

The concentration of nitrate in oceanwater is some ten orders of magnitude lower than the theoretical thermodynamic equilibrium between oxygen and nitrogen. This anomaly is related to the supply of oxidizable organic material from photosynthesis: many microorganisms in oxygen-deficient environments such as ocean sediments are able to use nitrate as an alternative electron acceptor to oxygen and in so doing liberate N_2 or N_2O to the atmosphere: the process of *denitrification*. On a geological time scale this has scavenged nitrate and maintained the high atmospheric dinitrogen content.

Sedimentary rocks, for this reason, contain little nitrogen and, because the element and its compounds are thermolabile, igneous rocks are also deficient. The consequence, for soil formation and plant succession, is that their early stages are governed by nitrogen availability through various and conflicting processes of microbial metabolism. The running tap of nitrogen fixation and

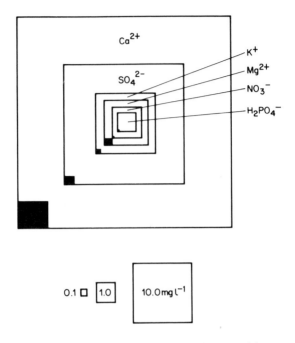

Fig. 5-1 Dissolved ions in wetland ground-waters. The areas of the open squares are proportional to the concentrations of the named ions in eutrophic waters and those of the shaded squares to concentrations in oligotrophic water. This presentation dramatically illustrates the enormous differences in concentration but even so the oligotrophic values for NO_3^- and $H_2PO_4^-$ have been exaggerated to make them visible. (Data of HASLER, 1975.)

the missing bath plug of denitrification, model the sources and sinks, while the primary producers depend on the regular recycling of the bath contents, the organic and inorganic compounds of nitrogen.

Nitrogen-fixing micro-organisms convert N_2 to organic $^-NH_2$ through the mediation of the molybdenum-based nitrogenase enzyme which is confined to a few free-living heterotrophic bacteria, some photosynthetic Cyanobacteria (blue-green algae) and a limited group of plant symbionts; bacteria of the *Rhizobium* group or Actinomycetes. Nitrogen fixation is an energy intensive process but the cost to the organism is repaid by competitive ability.

Some $^-NH_2$ may be leaked directly to plants, in particular from symbionts, but the major pathway of fixed nitrogen is via decomposer micro-organisms in which hydrolysis of proteins and deamination liberates NH_3 or NH_4^+ to soil solution. Many wetland plants absorb their nitrogen directly as NH_4^+ but in drier soils NO_3^- is the more normal source, the oxidations $NH_4^+ \rightarrow NO_2^- \rightarrow NO_3^-$ being, respectively, a source of energy for the chemolithotrophic *nitrifying* bacteria *Nitrosomonas* and *Nitrobacter*.

The microbial process of denitrification: $NO_3^- \rightarrow NO_2^- \rightarrow N_2O \rightarrow N_2$, previously noted as a function of anaerobic soils and sediments, is responsible for the thermodynamic non-equilibrium of N_2 and O_2 and for the active recycling of nitrogen from organic compounds via amminification and nitrification. Most soils in climates with $P/E > 1$ are automatic denitrifiers as they lack anion-exchanging sites and the NO_3^- ion is leached downward to less oxygenated soil (Fig. 5-2a). Wet soils above a water table also favour nitrogen fixation, nitrogenase being inhibited by too high a level of oxygen. Presumably the groups of micro-organisms involved in nitrogen-fixation, nitrification and denitrification are a well co-evolved interdependent microbial ecosystem, a supposition supported by their similar pH optima near neutrality. Only blue-green algae fix nitrogen actively in acid soils and, because nitrification is inhibited, NH_4^+ is the plant nitrogen source in these conditions.

Symbiotic *Rhizobium* spp. are less sensitive to soil conditions, the leghaemaglobin content of the root nodules serving to control O_2 supply and cellular metabolism providing a pH-stat. Despite this, some nitrogen-fixers on very wet soil, such as alders (*Alnus* spp.), form nodules above the surface amongst leaf litter. Possibly their actinomycete symbiont is more sensitive to soil oxygenation conditions. The association of nitrogen-fixers with the rhizosphere is also important, in some wet soils providing 10–20% of the plant nitrogen.

Soil crumbs (aggregates) may be surrounded by air space containing free oxygen but if the crumbs are fairly wet, limitation of gaseous diffusion allows their centres to become anoxic (Fig. 5-2b). In these conditions inward diffusion of nitrate again feeds a denitrifying system and loss from many well-drained agricultural soils is probably by this mechanism.

The nitrogen balance of rainfed peat soils must be dominated by contributions from blue-green algal fixation as they are too acid and calcium deficient to support most free-living bacterial fixers except perhaps for the tropical genus *Beijerinkia*. Blue-green algal nitrogen-fixation is also predominant in some forms of wetland rice culture, the high temperature and flooded soil encouraging both free-living species and also the symbiosis of *Nostoc* with the water fern (*Azolla*).

5.3 Carnivorous plants

Most of the world's wetlands support some form of carnivorous plants which trap insects and other small animals. Formerly it was believed that the prey was digested, as in the animal gut, and that the nutrition of the plant was improved by the process. Recent work with antibiotic and fungicicide applications to the traps suggests that the digestive process is microbiological and hence insectivory is a form of symbiosis, at least in the sundews (*Drosera* spp.). The use of tracer-labelled prey has confirmed that whole amino acids (or polypeptides) may be absorbed and experimental analyses of growth confirm the improved performance of plants supplied with nitrogen and phosphorus through this process. It seems reasonably certain that other elements such as calcium, deficient in acid peatland habitats, will also be supplied by carnivory.

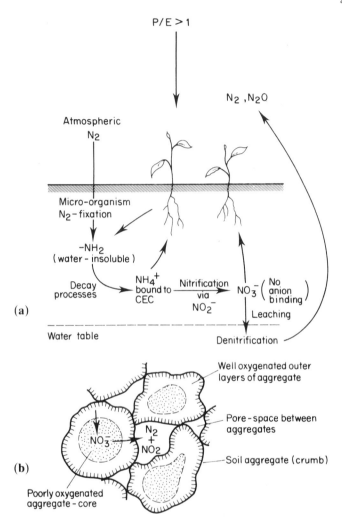

Fig. 5-2 (a) Most soils act as automatic denitrifying systems once mineralization has produced NO_2^- and NO_3^- which are not substantially bound by ion exchanging activity and are consequently leached to deeper, poorly oxygenated soil where they are denitrified. By contrast NH_2^-N is immobile because it is insoluble in water, other than in free amino acids, and NH_4^+ is bound by ion exchange to the cation exchange complex (CEC). P/E - ratio of precipitation to evaporation. (b) Denitrification in water-saturated soil aggregates in a rain-wetted but unflooded soil.

The trapping mechanisms, at their simplest are pitfall traps such as those of pitcher plants (*Nepenthes* spp.), perhaps baited with biochemical attractants. More complicated slowly-moving traps with inrolling leaf laminae and adhesive tentacles are found in the Droseraceae while the most advanced 'spring traps' of the Venus flytrap (*Dionea muscipula*) are able to trap active insects like dragonflies and fascinate children of all ages so that they are now popular house-plants. Beautiful photographs and details of carvivorous plant culture are given by SLACK (1979).

5.4 Flowing water and tidal replenishment

Many wet soil systems receive a constant flow of water which imports and exports soluble nutrients and also both oxygenates soil and removes soluble toxic products of the reducing environment. Freshwater and tidal environments are 'pulse stabilized', the seasonal or diurnal oscillations of water level being a regular physical and chemical perturbation which provides the energy for rapid nutrient cycling and, by disturbance, maintains the ecosystem in an early successional state, much of its primary production being deflected to the support of a very diverse consumer population. A fluctuating water table combined with differences of surface elevation also increases the number of niches for plant species hence the great biological diversity of these habitats.

Nutrient culture experiments with terrestrial plants has shown that essential elements are absorbed more efficiently from very dilute flow-replenished solutions than from static solutions of much higher concentration (ETHERINGTON, 1982). This, in itself, suggests that many wetlands might be net sinks for nutrients and some environmental measurements support this conclusion. Tidal water flowing through mangrove swamps may lose virtually all of its dissolved nitrogen and phosphorus, while in the very different peatland environment, substantial quantities of cations may be scavenged from through flowing water by ion exchange activity of *Sphagnum* mosses with consequent acidification of the outflow water.

Despite considerable research (HASLER, 1975; BURTON and LISS, 1976; GOOD, WHIGHAM and SIMPSON, 1978) it is still not known what conditions make wetlands net importers or net exporters of nutrients. Lakeshore and riverine reedswamp plants 'mine' nutrients from deep sediments and prevent their loss by burial or denitrification by depositing them in surface litter. In high energy tidal or flowing freshwater conditions much of this litter may be swept away but in the low energy lake-shore system it remains to be decomposed or accumulate as peat. The 'leaking' of nutrients in litterfall by individual plants in this case conserves nutrients within the ecosystem as a whole.

In flowing-water systems, import and export vary seasonally for some elements, thus analyses of marsh discharges show that plant uptake reduces the concentration of elements such as phosphorus, calcium and magnesium during spring and early summer but in late summer the situation reverses because a pulse of nutrient is released either by leaching of litter or its mineralization. By contrast nitrogen is nearly always depleted in outflowing water because of

Table 1 Nitrogen, phosphorus and calcium concentrations in river waters flowing from various vegetation zones in Uganda (Viner in HASLER, 1975).

Vegetation	Number of rivers	NO_3-N	mg l^{-1} PO_4-P	Ca^{2+}
Forest and wooded savanna	9	0.29	0.16	15.9
Steppe	4	0.20	0.25	8.4
Papyrus swamp (*Cyperus papyrus*)	11	0.02	0.01	4.7

denitrification in the marsh system. In tropical conditions there is evidence that discharge from swamps is depleted of phosphate, nitrate and calcium, suggesting that they act as nutrient sinks (Table 1).

Anaerobic soils themselves may act as traps for some elements. Where sulphide is produced, the very low solubility products of many transition element sulphides can result in their accumulation. The high concentrations of sulphur and heavy metals which are characteristic of some coals and their associated carbon-rich shales are consequent on such trapping in a swamp environment (Table 2). The interfaces between reducing and oxidizing layers of soils and sediments may be identified by precipitation of hydrous iron oxides: these also may scavenge heavy metals from solution by co-precipitation and also immobilize phosphorus by production of insoluble Fe^{3+} phosphates: re-oxidation of reducing sediments often produces a pulse of soluble phosphate, the solubility product of Fe^{2+} phosphate being more than a thousand times greater than that of the Fe^{3+} salt.

Table 2 Heavy metals and sulphides in sediments, soils, shale and coal.

	Pb	Zn	Cd	Cu	Co	Ni	Cr	Hg
			Element (μg g^{-1})					
River Elbe sediments (BURTON and LISS, 1976)	430	142	21	161	51	126	175	76
Average soil	35	90	0.21	30	8	50	70	0.6

	Oxidizable sulphides–S (%)
Sediments	0.2–0.65
Shales	to 7
Average coal	1.5
Mangrove soils	1–1.7

6 Wetland Productivity

6.1 Primary production: a global survey

Net primary production is conventionally defined as the difference between gross photosynthetic production and overall respiratory loss of dry matter (kg dry weight m^{-2} time^{-1}). It is ecologically significant as the maximum sustenance available to consumer organisms and, in this context, is better expressed as carbon per unit area (kg C m^{-2} time^{-1}) with the equivalence 1 kg plant dry matter $\equiv$ 0.5 kg C $\equiv$ 1.8 kg CO_2. In studies of energy flow, calorimetric measurements of energy content are used to convert mass flux to energy flux: plant tissues average 17 MJ kg^{-1} and animal tissues 24 MJ kg^{-1} (ETHERINGTON, 1982).

The productivity of ecosystems may also be described by the efficiency of conversion of solar radiation, for example in the temperate zone (Kew, England) mean daily flux of photosynthetically available radiation (PAR) is approximately 50 W m^{-2} (W = J s^{-1}; an annual input of almost 1600 MJ m^{-2}). A very productive agricultural grassland might produce 2 kg dry matter m^{-2} y^{-1} which, with an energy content of 17 MJ kg^{-1} represents a total annual fixation of 34 MJ m^{-2}, just exceeding an efficiency of 2% of the incoming radiation. This is near to the highest annual effiiency achieved by any ecosystem, wild or cultivated and most of the world's surface manifests much lower values because of limitation by cold, water deficit or shortage of essential elements. In total, the efficiency of the global ecosystem is only about 0.1% though it is obviously difficult to arrive at any critical estimate.

Wetland ecosystems are rarely droughted and often recieve supplementary nutrient inputs. It is not surprising then to find that the world's most productive wild vegetation is reedswamp (LIETH and WHITTAKER, 1975). It has been shown in previous chapters that wetland soils are hostile to roots and that the plants which are able to survive in them are herbaceous, mostly monocotyledons lacking secondary thickening. This lack is significant as it prevents the plants from accumulating non-photosynthesizing, respiring tissue, thus a given biomass of leaf canopy in the wetland environment is likely to have a higher net productivity than the same canopy in a terrestrial habitat where large volumes of secondary tissue have usually accrued (Fig. 6-1).

Because the habitat limits accumulation of living biomass, the annual production must be immediately recycled, maintaining the juvenility of the system, or it must accumulate as peat unless it is exported in high energy riverine or tidal conditions. Typically, wetlands produce more than they decompose and when the accumulation of peat or trapped sediment has raised the surface to the mean level of the water table it is common for woody species

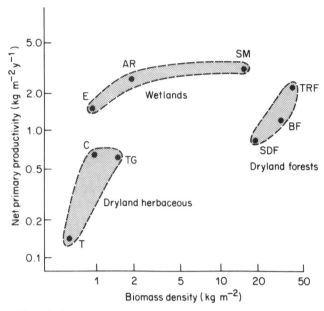

Fig. 6-1 The relationship between biomass density and net primary productivity of different vegetation types AR, algal beds and reefs; BF, boreal forest; C, crops; E, estuaries; SDF, deciduous forest; SM, salt marsh; T, tundra; TG, temperature grass; TRF, tropical rain forest. Note the logarithmic scales. (Data from LEITH and WHITTAKER, 1975).

to become established as in fen-carr or mangrove. As the habitat becomes drier, a larger proportion of each year's net production is diverted to storage as wood, less is available for the support of food chains and the ecosystem becomes more 'mature'.

6.2 Reedswamp, fen and marsh

Reedswamp plants such as the temperate common reed (*Phragmites australis*) or the tropical papyrus (*Cyperus papyrus*) have a rather variable ratio of shoot to underground rhizome and root dry weight, but typically the buried organs make up 25% to 50% of the total biomass and, because of their oxygen deficient environment, will probably have a lower than normal respiratory rate.

The maximum aerial biomass density (dry matter weight per unit area) of these plants ranges from 2–5 kg m^{-2} and it comprises largely photosynthetic tissue which is very efficiently exposed in a canopy structure of near vertical sword-like leaves with a low extinction coefficient of 0.3–0.4 (ETHERINGTON, 1982) which permits even the bottom leaves of a multilayered canopy to receive adequate light for some photosynthesis. The maximal photosynthetic rates of individual leaves are slightly greater, at 15–35 mg CO_2 dm^2 h^{-1}, than those of many terrestrial plants with Calvin-cycle (C–3) photosynthesis.

Reedswamps rarely suffer drought: in northern Europe their transpiration is near to the calculated potential value and ranges from 5–10 kg m^{-2} day^{-1} (Rychnovska in DYKYJOVA and KVÉT, 1978) which may be compared with values between about 0.5 and 2.0 kg m^{-2} day^{-1} for grassland under water stress. It may be inferred that water cannot be limiting and nutrient input in flowing water prevents mineral limitation. It is hardly surprising then that the maximum observed primary productivity of reedswamps, more than 6 kg m^{-2} y^{-1}, exceeds even that of cultivated ecosystems (LIETH and WHITTAKER, 1975) and must be close to the potential production calculated from radiant input, exceeding 5% efficiency of PAR conversion.

These very high productivity values are usually confined to monospecific stands of plants such as *Cyperus papyrus*. In drier sites, flooding stress is less severe, more species can coexist but the productivity is less; indeed there is a reciprocal relationship between wetland species-density and productivity. Furthermore, surfaces which are topographically or successionally drier become progressively invaded by terrestrially-adapted species which accumulate more non-photosynthetic tissue and very high photosynthetic efficiencies are no longer encountered. The early phases of marsh and fen are thus similar to reedswamp but as soil or peat accumulate, the productivity is likely to decline: an interesting difference between this situation and the conventional view of succession.

6.3 Mires

Ombrotrophic mires are rarely affected by drought but their mineral nutrition is dependent on wind-blown dust, solutes in precipitation and nitrogen-fixation is probably limited to blue-green algae. Again, it is hardly surprising that the contrast with the eutrophic wetlands in very low annual production rates. In northern England, carpets of the bogmoss *Sphagnum rubellum* produce about 0.13 kg m^{-2} y^{-1}: this is of the same order as tundra and desert ecosystems, represents an efficiency of c. 0.4% PAR but is still much higher than the global mean which is so depressed by the enormous nutrient-deserts of thermally stratified tropical ocean.

Greater productivity is found amongst the sub-aquatic bog mosses such as *S. cuspidatum* (0.5 kg m^{-2} y^{-1}) and also in drier areas colonized by heather (*Calluna vulgaris*) and cotton grass (*Eriophorum vaginatum*) which produce 0.66 kg m^{-2} y^{-1}, about as much as very poor quality deciduous woodland. In these acid conditions decomposition is slow, ranging from 0.04 kg kg dry weight^{-1} y^{-1} in aerobic peat, to a minute 10^{-8} kg kg^{-1} y^{-1} below the perched water table: hence peat accumulation!

Rheotrophic mires include the highly productive fens but even if the groundwater is acid and oligotrophic, it provides some nutrient input which increases productivity, species diversity and potential for agricultural reclamation.

Table 3 Primary productivity of coastal and freshwater ecosystems (data of LEITH & WHITTAKER, 1975 and CHAPMAN, 1977).

	Net primary production $(\text{kg m}^{-2}\,\text{y}^{-1})$	
North American salt-marsh tidal inundation)	3.9	calculated from energy flow data assuming 17 MJ kg^{-1} dry weight
North American salt-marsh	0.7	(infrequent tidal inundation)
Mangrove in Puerto Rico (*Rhizophora mangle*)	3.0	calculated from daily estimate assuming all year growth
Seagrass communities (e.g. *Zosteraceae*; *Cymodocaceae*)	4.2	
Freshwater swamp maximum	6	

(For comparison; agriculture maximum 4)

6.4 Coastal ecosystems

The tidal pulsing of salt-marshes and mangroves subsidizes the nutrient cycle and maintains juvenility by constantly resetting the successional clock. They are of high productivity but do not out-yield the most fertile freshwater systems because substantial amounts of energy must be used in solute-control by manufacture of osmotically-balancing biochemicals, direct ion-pumping as in salt glands or production of succulent tissue with its high relative respiratory cost. Some fully-submerged coastal systems such as the seagrasses (e.g. *Zostera* and *Thalassia* spp.) do however out-yield croplands. Examples are given in Table 3.

6.5 Agriculture and utilization of wetland biota

Rice is the most significant wetland crop both in terms of world production and impact on man, probably more than half of our population depending on it as a staple food. Rice is the highest yielding grain crop, growing in hot climates and fed with water by irrigation or monsoonal flooding: worldwide its total production is second only to wheat (Table 4). The breeding of short-straw 'green revolution' varieties permits large grain-yield responses to nitrogen phosphorous potassium fertilization, without increased height growth and wind damage by lodging, but the dependence of nitrogen-fertilizer economics on oil price has proved an unforseen problem in the full exploitation of these varieties.

Next to rice cropping, the most widespread use of wetlands is for grazing domesticated animals and, with little or no management, for production of game birds, fur-bearing mammals and as artificial fishponds in which carp and other species depend on reedswamp and shallow-water primary production. Freshwater grazing lands may pose problems of parasitism, for example the sheep liver fluke (*Fasciola hepatica*) which is transmitted by the water snail *Lymnaea truncatula*. Saline tidal grazing are free of this particular difficulty and, in northern Europe, (*Spartina anglica*) cord grass provides fairly high grade cattle and sheep pasturage (Table 4).

Traditionally, in temperate countries thatch roofing has been a wetland product, mainly from reedswamp plants, though asbestos-cement and galvanized steel have now largely ousted it even from poor rural communities. In Europe, the common reed (*Phragmites australis*) and giant reed (*Arundo donax*) have been the most widely used, the latter also for fencing and windbreaks. The leathery leaves of *Spartina alterniflora* (cord grass) from North American salt-marshes also provide a low grade thatch. Basketwork is another craft relying on reed production and also on plaint willow stems cultured in marsh and carr osier beds (*Salix viminalis*).

Many rushes (*Juncus* spp.) and other reed-like plants found domestic use as a strewn floor-covering and before the advent of oil and paraffin-wax candles, rush-lights were manufactured from the porous stem pith of *Juncus effusus* by soaking in molten animal fat. In Britain, the candle tax of 1709 caused a resurgence of the rush-light in poorer homes and their use continued until the nineteenth century in Wales and Ireland.

The wetter parts of European and North American salt-marshes have traditionally been used for shellfish culture, taking advantage of the rich detrital foodchains supported by the upper marsh. Oyster yields of 5000 kg ha^{-1} y^{-1} have been achieved in North America (CHAPMAN, 1977) while the game-fisheries of Florida and many shrimp fisheries likewise depend on exported litter from mangrove or salt-marsh.

Table 4   Commercial yields from wetlands.

Rice (*Oryza sativa*) grain yield (t ha^{-1})	
Average	5–6 (intensive agriculture)
Record	10.6
Rice: 1970 world total (t)	2.75×10^8
(Wheat	2.98×10^8)
Salt-marsh pasture (*Spartina anglica*) (t ha^{-1})	8–12
Reeds(*Phragmites australis*) Czechoslovakia (Commercial bundles 1.5 m × 1 m circumference ha^{-1})	350–1000

The cord grasses (*Spartina* spp.) have proved useful in coastal reclamation and stabilization of dredge-spoil. They may be established from seed in mid-marsh conditions or hand-planted as rhizome cuttings at all levels of tidal flooding. In northern Europe the ease with which *S. anglica* becomes established has allowed it to invade a wide range of North Sea and Atlantic coastal marshes where its rapid colonization rate has made it an ecological management problems. In the Netherlands, common reed (*Phragmites australis*) has also been used in the early phase of colonization and desalination of drained polders.

Wetlands have considerable ability to 'sink' polluntants of all kinds including nitrogen and phosphorus from sewage and agricultural drainage, heavy metals and artifical products such as chlorinated hydrocarbons. For this reason, wetlands and in particular coastal salt marshes have found use for wastewater disposal. CHAPMAN (1977) cites their annual value for this purpose as several times that of an alternative use such as oyster culture. Many ecologists are, however, worried not only about the duration of the sink activity but also the biological consequences for marshes of scientific value.

Wetlands, in agriculture, are generally of nuisance value, waterlogging both damaging crops and impeding tillage. A complex industry has grown from the traditional farming operation of ditching: sectional clay pipe (tile), rubble and mole drains were introduced during the last century and now replaced by machine-installed, continuous plastic pipe drains.

It is beyond the scope of this text to discuss the hydrological theory of soil drainage but it is interesting to note a few of its consequences. Discharges from oxidized FeS-containing soils may contain both H_2SO_4 and colloidal iron oxides. The former may cause fish-kills in watercourses and the latter often gives problems of drain-pipe blockage with ochre (hydrated iron oxide). Drainage of heavily fertilized land intercepts NO_3^- which would otherwise be denitrified and may cause eutrophication of streams and rivers.

7 Consumption, Decomposition and Elemental Cycling

7.1 Comminution of litter

When leaves and branches fall to the ground and roots die, the beginning of the decomposition process is strongly dominated by animals which eat, shred or bury litter and return faecal pellets to the soil. This first processing of dead organic remains is mainly a physical comminution which opens the door to chemical change by micro-organism attack and to colonization by micro-arthropods, nematodes, protozoa and other organisms which consume either micro-organisms or comminuted litter.

In fertile terrestrial soils most litter is eaten or buried by the larger lumbricid earthworms or, in some tropical habitats, by termites. Absence of oxygen excludes many of these from waterlogged environments, though the smaller enchytraeid worms occur in some quantity in wet organic soils and, despite their size, have a relatively much higher metabolic rate and contribute substantially to litter transformation, particularly in acid habitats.

Little seems to be known of comparative energy-flow paths through these detrivores of different soils but it is likely, in the absence of active earthworms, that their place as primary processors and buriers of litter will be taken by molluscs, dipterous larvae, coleoptera and their larvae, isopod and amphipod crustacea. In very wet leaf litter of forest pools, springs and streamsides, saprophagous isopods and amphipods may be the most abundant comminuters, for example the water lice (*Asellus* spp.) and freshwater shrimps (*Gammarus* spp.). Gammarid amphipods and the marine isopod *Cleantis planicauda* are the most active comminuters of *Spartina alterniflora* (cord grass) in North American salt-marshes while amphipods and crabs fill a similar role in mangroves.

7.2 Microbiology of wet soils

The invasion of litter by fungi and bacteria and its comminuted products marks the commencement of mineralization and the closure of the organic-inorganic circle. Under aerobic conditions the process is well known but in waterlogged anaerobic soils and in the oxygen-deficient centres of soil aggregates, a less familiar microbiology is encountered.

Over 90% of the carbon dioxide evolved from soils is generated by micro-organism respiration and, in many dryland soils, a large proportion of this is attributable to filamentous fungi. These are almost all strict aerobes and are eliminated by waterlogging as are the obligately aerobic bacteria such as

Pseudomonas spp. Their place is taken by a range of facultative or obligate anaerobes and it is to these that wetland soils owe their distinctive redox chemistry.

The elimination of aerobes from fully-waterlogged soil limits the rate of decomposition, possibly initiating peat-formation and certainly doing so when lack of oxygen is compounded by low pH and shortage of Ca^{2+}. The decomposition which does occur in the aquatic environment depends on the slow inward diffusion of oxygen to satisfy the electron acceptor requirement of normal respiration or on the presence of other electron acceptors and their regeneration, as detailed below. Alternatively, many anaerobes have a fermentative metabolism, thus many Clostridia provide themselves with energy by butyric fermentation, but, at the same time, leak organic products such as butyric acid to the environment, adding to its phytotoxicity and biological oxygen demand.

Many bacteria, in the absence of oxygen, begin nitrate respiration: nitrate provides an oxygen source and is itself reduced to nitrous oxide or molecular nitrogen which are lost from the soil in the process of denitrification: (§ 5.2).

In waterlogged soils, species of *Achromobacter* and *Bacillus* are active denitrifiers but the process is strongly inhibited by oxygen, even 0.1 to 0.2% v/v reducing the rate to about one tenth that of an anaerobic system. Aerobic *Pseudomonas* spp. also denitrify under low oxygen partial pressures. Despite this activity, waterlogged soils are not necessarily deficient in plant-available nitrogen because free-living anaerobic bacteria such as *Clostridium pasteurianum* and *Desulphovibrio* spp., Cyanobacteria (blue-green algae) in addition to facultative anaerobes (e.g. *Bacillus polymixa*) are nitrogen fixers. Breakdown of organic matter in these soils produces NH_4^+ which is assimilated by wetland plants and is only denitrified rapidly if an oscillating water table permits its oxidation to nitrate through the normal nitrification pathway.

After flooding, oxygen is scavenged from the soil, nitrate reduction begins and at its limit the redox potential will have fallen to about $E_7 = 250$ mV (Fig. 2-2). At this potential Mn^{4+} and Mn^{3+} act as electron acceptors and the much more soluble reduced Mn^{2+} appears in solution; for this reason it is unusual to find nitrate and Mn^{2+} in the same sample of soil water. It is still not clear whether manganese reduction is simple a concomitant of falling redox potential and changing pH, or whether it is bacterially mediated. Several bacteria including *Metallogenium* spp. are responsible for Mn^{2+} oxidation when oxygen or nitrate is available and may cause the formation of concretions of black manganese dioxide (MnO_2) in some gleyed soils.

The sequence of changes shown in Fig. 2-2 indicates that reduction of manganese is followed by that of iron. Here there is more knowledge of the involvement of heterotrophic bacteria, thus has shown that some obligately anaerobic *Clostridium* spp. reduce iron, and postulated an unknown reductase enzyme which catalyses the acceptance of H^+ ions by iron oxide with the release of Fe^{2+}. Gley formation was simplistically interpreted as conversion of brown $Fe(OH)_3$ to green-grey $Fe_3(OH)_8$.

When the oxygen supply is restored by a falling water table, by diffusion through root systems or when Fe^{2+} is carried onto an oxidizing environment by percolating water, oxidation occurs either inorganically or as an energy-yielding metabolic process of autotrophic iron bacteria. In the very acid environment associated with oxidizing sulphides (section 2.2) this may be the bacterium *Thiobacillus (Ferrobacillus) ferrooxidans* but in soils and water nearer to physiological pH ranges, the stalked bacterium *Gallionella* and filamentous bacteria of the *Sphaerotillus* group are involved. The Fe^{2+} ion acts as an electron donor, O_2 as an acceptor and the Fe^{3+} which is produced, initially precipitates as gelatinous flocs of $Fe(OH)_3$ which then alter with time to form the hydrated oxide flecks so typical of gleyed soils. Bacterial activity was largely responsible for the deposits of bog iron ore used by primitive man, and probably for most sedimentary iron-ore deposits later than Pre-Cambrian.

With duration of flooding, or depth from an oxidizing interface, various organisms exploit the redox ladder of chemical changes and once iron has been reduced the next step involves the conversion of SO_4^{2-} to H_2S, mainly by *Desulphovibrio desulphuricans* and *Desulphotomaculum* spp. which utilize H_2 or organic materials as an electron donor. The formation of sulphide in the presence of Fe^{2+} causes the chemical precipitation of FeS already described but it also provides an electron source for a wide range of sulphur bacteria if a suitable oxidant becomes available. The photosynthetic purple and green sulphur bacteria are, for obvious reasons, not common in soil but occur very extensively in shallow water overlying reducing sediments such as tidal mud-flats. The purple bacteria are strongly halophilic and may impart a strong red-pink coloration to salt evaporation pans such as those of the French Camargue and of San Francisco Bay.

Very reducing soils and sediments liberate both methane and hydrogen as 'marsh gas'. It is inflammable, as Volta first demonstrated in 1776 using his 'electric pistol' eudiometer and its ignition, perhaps by spontaneous combustion of phosphine (PH_3), may be the cause of the legendary 'Will o' the wisp'. Many morphologically diverse anaerobic bacteria, for example *Methanobacterium* and *Methanomonas* spp., generate methane by reducing CO_2 with H_2 as an electron donor. In the same reducing sediments butyric fermentation of carbohydrate, particularly by *Clostridium* spp., liberates H_2 and CO_2 from pyruvate.

Upwardly diffusing hydrogen and methane are intercepted by micro-organisms when they reach oxygenated soil or when they meet oxidizing compounds. Hydrogen thus donates electrons to oxygen, sustaining *Hydrogenomonas* spp., or to sulphate in the presence of sulphur-reducing bacteria to produce H_2S. Some bacteria reduce CO_2 with H_2, thus *C. aceticum* produces acetic acid in this way while methane is oxidized by bacteria such as *Bacillus methanicus*.

7.3 Elemental cycles

The elemental cycles of wetland environments differ from those of terrestrial habitats in the ability of flowing water to import or export particulate and dissolved materials and also in the modification of decomposition and recycling processes by the presence of low redox soil or sediment which causes loss of nitrogen but trapping of carbon, sulphur and some heavy metals. The general nature of elemental cycles has been discussed in this series by JACKSON and RAW (1966) and ETHERINGTON (1982): no further information is presented here other than that related to the specialized habitat (Fig. 7-1).

The annual nitrogen uptake of wetland vegetation is similar to that of other productive habitats as the elemental content of herbaceous tissues differs little between wet and dry ecosystems (Table 5). In forests, despite woody tissues containing much lower concentrations of many elements, the annual cycle of uptake and litter return is dominated by leaf-fall, thus the main quantitative difference between the nitrogen cycle of wetlands and terrestrial habitats is related to dentrification, failure of mineralization and transport by flowing water.

The denitrification rate of eutrophic wetlands is potentially enormous, in particular when they are enriched by sewage effluent or fertilizer (Table 6). Even without fertilization, denitrification rates are of the same order as those of agricultural land and as dryland denitrification is mainly confined to deeper, poorly oxygenated soil, it is reasonable to assume that most of the global denitrification is a function of wet soils and submerged sediments: about one third of earth's primary productivity is marine, and yet half the denitrification takes place in continental shelf water, much of it being nitrogen originally fixed on the land surface. Denitrification in acid soils and peat is less active and in such habitats relatively large amounts of immobile organic nitrogen may accumulate.

Import and export by moving water has a substantial effect on the nitrogen cycle, for example in tidal mangrove swamps about half of the 0.5 kg m^{-2} y^{-1} leaf litter is washed out to nearshore waters. *Rhizophora mangle* leaves contain about 1% nitrogen, equivalent to a nitrogen loss of 25 kg ha^{-1} y^{-1}. The phosphorous content of the litter is about 0.2%, representing a loss of 5 kg ha^{-1} y^{-1}. Odum and Heald (in HASLER 1975) have suggested that this exported organic matter supports a detrivore food-chain which terminates in important food or game fisheries.

In some instances, seasonal flooding brings the detrivores to the litter rather than *vice versa*. An interesting example is the months-long flooding of Amazonian swamp forest (igapo) of *Eugenia inundata* in which litter provides fish with an unusual means of support and thereby a protein supplement to the diet for humans.

The trapping of sulphur by anaerobic sediments is a major biogeochemical process, operating through microbiological sulphate-reduction and precipitation of FeS in the presence of Fe^{2+}. In constantly anaerobic conditions there is no mechanism for release of this very insoluble sulphide and, over long time periods, diagenesis to pyrites (FeS$_2$) occurs and is responsible for the very high sulphide content of coal and many sedimentary black shales (Table 2).

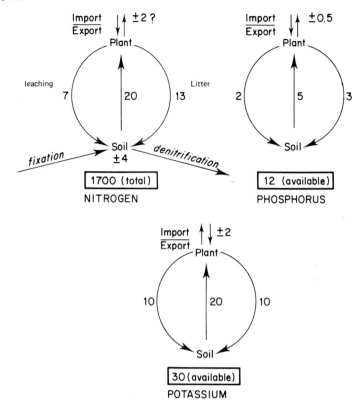

Fig. 7-1 Nitrogen, phosphorus and potassium cycling in a eutrophic wetland. The compartments represent soil resources of each element $(g\,m^{-2})$: for nitrogen this is total organo-N, for phosphorus; acidified ammonium fluoride extractable and for potassium; carbon exchangeable. Arrows indicate transfers $(g\,m^{-2}\,y^{-1})$. (Data based on Klopatek in GOOD *et al.*, 1978), except for import:export and nitrogen-balance figures which are estimates based on general literature.

Table 5 Percentage elemental content of wetland and other vegetation.

	Wetland herbaceous vegetation (%)	Terrestrial grassland	Deciduous trees: woody tissue
N	1–4	1–3.5	0.5–1.5
P	0.1–0.6	0.1–0.4	0.01–0.06
K	0.4–4.0	0.4–4.3	0.07–0.3
Ca	0.2–0.8	0.1–2.5	0.2–1.1

Data of Boyd in GOOD *et al.* (1975) and others.

Table 6   Dentrification in wetlands and agricultural ecosystems.

	$kg\,N\,ha^{-1}\,y^{-1}$
Salt-marsh (North America)	50
Salt-marsh plus 76 kg ha^{-1} N	100
Intensive agriculture	*33*
(San Joaquin Valley, California)	$kg\,N\,ha^{-1}\,d^{-1}$
Freshwater marsh	3.5
plus sewage effluent	

8 Peat

8.1 The time machine

For reasons already outlined, peat accumulates over any soil which is wet enough to inhibit microbial oxidation, particularly if it is acid and calcium deficient. The peat accumulates in a year by year sequence and because decomposition, at least in acid conditions, is very slow, many thousands of years of seasonal history may be preserved in the 'archives of the peat bogs' (GODWIN, 1981). Not only the peat-forming organisms and local biota are preserved but also materials blown or carried onto the peat surface: pollen and archaeological artefacts.

The former are preserved particularly well, the pollen grain being coated with sporopollenin, a very stable part-esterified carotenoid co-polymer. Different plant species are identifiable by the shape and surface sculpturing of the grains (Fig. 2-4) thus, by extracting them from peat (MOORE and WEBB, 1978) and counting proportions it is possible to comment on the prevailing vegetation type at the time of formation while ^{14}C-dating of buried organic materials produce an absolute chronology for such pollen analysis. The Quaternary palaeoecologist is thus afforded a privilege denied those who study older hard-rock fossil assemblages: the presence of a continuous sub-fossil record.

Analysis of pollen and macrofossils from peats all over the world have now established fairly detailed pictures of postglacial (c. 12 000 B.C. onward) vegetational change under the influence of climatic alteration and human impact (DIMBLEBY, 1977 in this series) while, in the tropics, pollen analysis has given new insight to the effect of the wetter and drier periods which have caused major advances and recession of tropical forest during the Quaternary period. In a few areas of North America and Eurasia which escaped the effects of the Devensian (last) glaciation, pollen and macrofossils in peat and lake sediments have given a rare view of conditions during the previous interglacials which terminated, respectively, 70, 140 and 275 thousand years ago.

Pollen analyses are best appreciated when presented in the form of chronological diagrams relatable to both absolute ^{14}C-dates and to identifiable prehistorical or historical events (Fig. 8-1). In this case data have been plotted for just a few species or groups which indicate very well the warming of the immediately post-glacial climate in Britain, followed by the rising and continuing impact of man to the present day.

Studies of both macrofossil and pollen content of peat may also be presented as stratigraphic diagrams (Fig. 8-2), often particularly useful in interpreting data from archaeological sites where human activity itself has been 'fossilized'

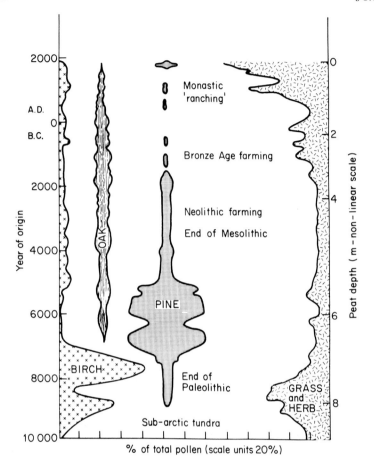

Fig. 8-1 Pollen diagram of bog in Co. Tipperary, Eire based on data of Mitchell cited in WEST (1968). For clarity only three named species and the total of grass + herb pollen have been plotted. The time scale extends over 12000 years and the diagram is annotated with details of environmental conditions and human activity. The depth scale shows that peat has accumulated at *c.* 0.75 m per 1000 years, commencing in the immediately post-glacial period. With amelioration of the climate the early conditions of grass-sedge and grass-birch tundra gave way to increasing colonization by birch (*Betula* spp.) followed by pine (probably *Pinus sylvestris*). The oaks (*Quercus robur* and *Q. petraea*) first appeared in about 7000 B.C. persisting to the present day, the last 2000 years of decline being accompanied by a sharp rise in grass-herb pollen as disafforestation increased.

by immersion in peat, sometimes to the extent of preserving man himself in such good condition that studies of tissues, gut contents and even clothing are possible (GLOB, 1969).

8.2 Succession revisited

Most speculation concerning succession relies on observable gradients of species composition such as those accompanying gradation of water level at a lake shore or a time sequence of surfaces exposed by glacial recession, silt or sand accumulation. Only rarely is it possible to study sub-fossil sequences at one spot and these are always related to the preservative characteristics of anaerobic sediments or peat.

Figure 8-2 shows direct evidence of succession from open salt-marsh mudflat, through a mixed salt-marsh vegetation followed by upper-marsh *Juncus maritimus* (sea-rush) invasion to alder-willow woodland on a site now dominated by *Salix atrocinerea* (grey willow) (RANWELL, 1972). WALKER'S (1970) classic refutation of the unidirectional hydrosere theory was likewise based on the study of macrofossil remains in peat-or sediment-filled basins and gave evidence for a wide variety of transitions between vegetation stages which had previously considered to be a linear time-sequence from open water to carr woodland or bog.

8.3 An exploitable reserve

Fossil carbon is a valuable commodity as we too clearly realize from the dependence of world economy on oil and prevously on coal. The earth's one to two million square kilometres of peatland despite having provided one of the earliest fossil fuels now makes but small contribution to the power needs of the developed world except for the U.S.S.R. which takes over 95% of the world annual harvest of peat. Of this annual production, exceeding 80×10^6 t, almost three-quarters is used for electricity generation. Apart from Russia, Eire is the only other country substantially using peat for power generation: in other countries the major use is as a horticultural soil ameliorant and compost base (MOORE and BELLAMY, 1973).

8.4 Forestry and agriculture on peat

Deep ombrotrophic peats are rarely reclaimable being too nutrient deficient, acid and toxic to support demanding crop plants or even conifers which are adapted to oligotrophic soils. Furthermore the problems of drainage, access with heavy machinery and potential wastage by erosion and oxidation are usually sufficient to dissuade the farmer or forester.

There are a few exceptions to this generalization, for example in North America the cranberry and blueberry (*Vaccinium* spp.) industry is based on such soils while in Britain attempts have been made to afforest deep peats with

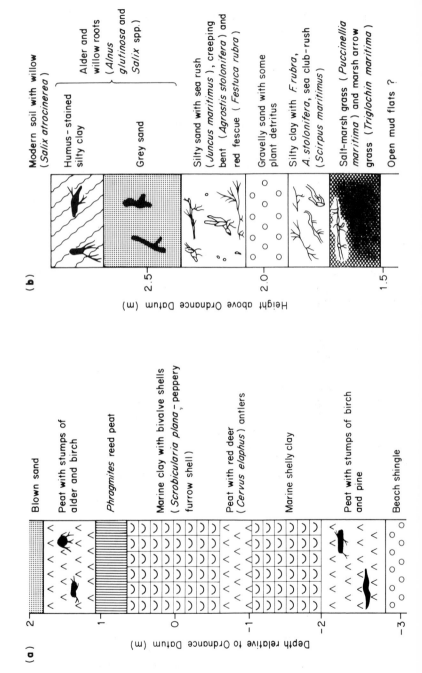

tolerant conifers such as *Pinus contorta* (lodgepole pine) and *Picea sitchensis* (sitka spruce). The former is claimed to dry-out peat deposits by its deep rooting habit but, in general, expensive drainage works are needed together with an initial phosphate fertilization. Most commonly ombrotrophic peats are used for rough grazing and game-bird production, either without management or with rotational burning to promote young shoot growth.

Rheotrophic peats, by contrast, have been widely reclaimed, the fens of East Anglian Britain being the classic example, but all over the world from the Everglades to the *chinampas* of Mexico, these nutrient-rich peats have been cultivated as remarkably rich farmlands, their only problems being the need for drainage and long term oxidation loss. A few shallow ombrotrophic peats overlying base rich deposits have also been exploited in this way by mixing peat with the underlying material.

Fig. 8-2 Stratigraphic diagrams of wetland peats and sediments. (**a**) Coastal peat-clay 'Forest Bed' from Swansea Bay, South Wales. Interlamination of peats with marine clays sea-level oscillation during the post-glacial period. Seawater inundation followed by gradual drying-out is shown well in the sequence of clay to *Phragmites* reedswamp followed by establishment of alder-birch woodland (0-1.75 O.D.). Pine in the deeper peat indicates a probable age exceeding 7000 B.P. (**b**) A salt marsh deposit showing succession from open mudflats to alder woodland cover in the Fal estuary, Cornwall (RANWELL, 1972).

9 Management of Wetlands

9.1 Hydrology and water quality

Wetlands are more vulnerable to accidental and intentional human interference than any other habitat, firstly because regional water level is so easily altered and secondly because some wetland plant communities are differentiated from each other solely by differences in water table of no more than one or two centimetres. A relatively small interference caused by abstraction of water, drainage, or civil-engineering works may thus have enormously magnified biological consequences. Tidal wetlands are equally at risk from coastal defences, impoundments or dumping of dredge-spoil which alter hydrology and may eliminate tidal pulse-stabilization (section 5.4).

A third reason for the fragility of wetlands is the ease with which water quality may be changed, sometimes by a single inflow to a basin system. The most striking example is eutrophication with additional sewage-borne phosphate or fertilizer derived nitrate which may cause 'blooming' of a few very productive species of algae, floating aquatics such as the tropical water fern and water hyacinth (*Salvina molesta* and *Eichhornia crassipes*), submerged aquatics or reedswamp species. Such blooms are followed by excessive decomposer oxygen demand, sulphide generation and fish death. The increased productivity also increases the rate of successional infilling of open water. Nutrients may not be the only water-borne pollutants, thus organic matter in sewage, or from paper making and other biologically-based industries, causes oxygenation problems while a host of toxic materials ranging from heavy metals to chlorinated hydrocarbons can cause plant and animal deaths or, by transfer in food chains, unexpected and horrific consequences for man.

9.2 Management for conservation

The conservation of wetlands is less compatible with other human activity than that of most other habitat types. When a reserve is designated, the continuing supply of water and its maintained quality must be guaranteed against subsequent demands for water supplies and for disposal of drainage which may be eutrophic or even toxic. Such guarantees, furthermore, may have to include large areas outside the reserve and thus need both legislation and compensation.

Public access and agriculture must be stringently restricted since wetland soils and vegetation are very easily destroyed by trampling and churning, the 'healing' process often long delayed by the creation of sunken, waterlogged tracks and patches. Where open water provides a magnet for fishermen, sportsmen and children, such restraints may be seen as conflicting with public ownership, a problem only soluble by forethought in resource planning and education.

The human-imposed problems of wetland reserves are exacerbated by the simple ecological fact of succession: wetlands are transient habitats unless they are pulse stabilized by tidal flooding, seasonal inundation, the wandering of an erosive river or repeated fire. A fence and time may convert the finest pond into a willow-scrub: maintenance of the status quo consequently requires radical interference with the water table (damming), soil surface level (bulldozers!) or vegetation and peat cover (axe and fire) operations which often find no favour with the public or even many reserve managers. The aesthetic impact of such work may be reduced by a systematic rotation in which only small areas are 'damaged' each year. This has the added advantage of creating a mosaic of different age habitats, particularly useful if some plant species are associated only with single successional stages. The increase in feeding-niche diversity may support additional animal species.

Another crucial decision at the time of designation is the size of reserve. In the context of 'island ecology' the reservation must be sufficiently large to reduce species extinction and migratory losses to an acceptable level. The wetland environment is particularly demanding, even a small basin mire or fen being supported by a much larger catchment while relatively modest river systems may drain many thousands of square kilometres. Wetlands which support migratory animals, particularly birds, can form part of an international network for which conservation measures must be co-ordinated, sometimes continents apart. The suggested tidal barrages in Britain: Morecambe Bay, the Wash, the Severn and Dee estuaries, may permanently flood the tidal mud feeding grounds of winter migrant birds such as knot (*Calidris canutus*) which breed as far away as northern Siberia. Many more species may suffer unless international co-operation limits such erosion of habitat.

From the early 1960s attempts were made to define criteria by which important wetlands could be identified as a preliminary to establishing an international legal framework for their protection. The International Union for Conservation or Nature (IUCN) and the United Nations Environment Programme (UNEP) *Directory of Western Palearctic Wetlands* (1980) presents a brief history of these developments. Project MAR, organized by IUCN, the International Council for Bird Preservation (ICBP) and the International Wildfowl Research Bureau (IWRB), during the early 1960s stressed the need for an international convention on wetland preservation. Within a decade the *Convention of Wetlands of International Importance especially as Wildfowl Habitats* was adopted at Ramsar, Iran in 1971, the United Nations Educational and Cultural Organisation (UNESCO) acting as a clearing-house for 29 countries which had signed the Ramsar Convention by 1980.

Simultaneously with these developments, Project AQUA, associated with the International Biological Programme (IBP), was responsible for drawing-up the AQUA list of lakes and rivers of high limnological interest: and in 1974 the

Heiligenhafen International Conference on the Conservation of Wetlands adopted four main criteria for the definition of important sites: these are outlined in IUCN-UNEP (1980):

   (*i*) Importance for waterfowl species and populations or for other threatened and important plant or animal species.
   (*ii*) Representative character or uniqueness.
   (*iii*) Research, educational or recreational value with particular stress on the first two.
   (*iv*) Practicability of conservation management.

The publication of the *World Conservation Strategy* (IUCN, UNEP and World Wildlife Fund, 1980) was an attempt to draw governmental attention to the urgency of conserving for sustainable development of earth's habitats. The objectives and requirements of conservation are outlined in the Strategy with priorities for national and international action. Wetlands are repeatedly mentioned in the context of protection against misuse and over-exploitation, the need to protect diversity with particular priorities for watershed management and the solution to the problems of fragile swampland coastal ecosystems.

Peatlands have not been neglected in the international drive for wetland protection. Project TELMA, initiated by IUCN and IBP is intended to provide a scientific basis for peatland conservation, a matter of extreme priority as the rates of destruction are enormous. Finland has already cultivated or afforested nearly four million of her 10–11 million hectares of peat while the Netherlands has totally lost all of an original half-million hectares. Added to these direct losses is the threat of eutrophication by groundwater contamination or atmospheric input in industrialized countries.

9.3 Regional management

Even today wetlands are too commonly considered as marshy wastes fit only for reclamation or dumping of materials. This attitude has caused enormous losses, though to the advantage of agriculture. The fenlands of East Anglia are now some of Britain's most fertile arable land; the French Camargue, once an impenetrable tract of reedswamps and salt-marsh is now a rich granary of rice, maize and other cereals; Dutch agricultural economy rests on reclaimed peats, tidelands and estuaries while the Florida Everglades, an enormously diverse sub-tropical wetland has lost perhaps a quarter of its area to cultivation and grazing during the past century.

The extent of the wildlife losses in British Fenland is vividly described by GODWIN (1978). The original vegetation was an amazingly rich apposition of acid raised bog, calcareous marl rivers and meres, fenland, fen-carr and fen-woodland of all stages, concealing a buried treasury of post glacial geology, archaeology and fossil plant remains. Early peat cutting and later drainage has removed a huge tract of peat over three metres in thickness, much of the peat wastage being caused by oxidation after drainage.

The loss of the peat has reduced the landscape to a geometrical pattern of arable and grassland fields, separated by drainage cuts. Most of the originally diverse flora and fauna has gone, with the exception of a few renowned nature reserves while the last 150 years has seen the extinction of the raised bog acid peat species.

The magnitude of wetland management problems may be illustrated in the nearby Broadlands of Norfolk. The river widenings, locally named Broads, are now known to be the remains of medieval peat-cuttings and closely resemble the lost fenland marl-lakes. By the eighteenth century they had become very rich wildlife habitats in which succession to alderwood was controlled by cropping of reed (*Phragmites australis*) and sword-sedge (*Cladium mariscus*) for thatch and by the dredging and cutting of aquatic macrophytes to maintain navigable channels.

During the nineteenth century the city of Norwich grew and the river Yare began to be sewage-eutrophicated. Originally clear-water marl-depositing rivers with a moderate growth of bottom macrophytes became blanketed with filamentous algae such as *Cladophora* spp. and, in the early 1900s when main sewerage was installed, organic pollution became apparent and eutrophication further increased. The waters became turbid with phytoplankton which shaded-out submerged macrophytes and reedswamp species were encouraged.

Post-War agricultural fertilization increased nitrate levels and domestic sewage became phosphate enriched from detergents. Loss of bottom macrophytes was finally complete and the sediments were no longer stabilized by plants.

During this same period from 1900 onwards the Broads developed a thriving holiday industry based on great fleets of hire-boats which not only increase the sewage problem but also disturb bottom sediments as well as rooted and reedswamp vegetation. The Broads have traditionally been a rich coarse fishery, now much exploited by holiday makers, and the loss of lead-shot fishing weights in shallow and reedswamp water is producing an increasing problem of lead toxicity for waterfowl, particularly bottomfeeders such as mute swams (*Cygnus olor*).

Eutrophication has other sources: Hickling Broad owes much of its problem to a black-headed gull (*Lapus ridibundus*) colony which has increased ten-fold in three decades, supported by the invertebrates of an increasing area of ploughland and urban rubbish tips. The birds are no longer attracted to the coast because the inshore fishery has declined. Each bird may contribute up to 40 mg of faecal phosphorus per day to the water!

Eutrophication has caused fish and bird deaths through the microbiological production of toxins. The brackish-water alga, *Prymnesium parvum* caused 250 000 fish to die in Hickling Broad during 1969, while birds have died from the botulism secreted by the anaerobic bacterium *Clostridium botulinum* which now grows on deoxygenated sediments in broads which have become organic-rich.

The extension of high productivity arable and grassland has been achieved by increasing deep drainage and lowering of regional watertables. Sulphide-containing soils, in these conditions, may generate sulphuric acid and ferrous sulphate which appears in drainage water as extreme acidification and blankets of gelatinous ferric hydroxide (section 2.2). In November 1970 the waters of Calthorpe Broad suddenly changed from c. pH 7 to c. pH 3, killing all fish and freshwater mussels along with most macrophytes. The acid-water pulse has followed autumn rain in most subsequent years, but from 1978 lime has been successfully used to control pH. Similar problems have occurred in Martham Broad and have prompted promises of governmental enquiries into the relationship between autonomous Internal Drainage Boards and Water Authorities which are legally responsible for water pollution control.

The decline of reedswamp, an important habitat for wildfowl and for food-plants of threatened species such as the swallow-tail butterfly (*Papilio machaon*) has been attributed to boating damage, eutrophication and successional loss due to cessation of thatch-cutting but the introduced coypu (*Myocastor coypus*), feral in the Broads since the 1930s has also caused serious direct damage by underwater grazing in winter and spring. Removal of leaves also makes the rhizomes susceptible to anoxia in eutrophic sediments in which they would normally be undamaged.

The present status of the Broadlands with a host of interacting manmade problems indicates the diffiiculty of managing this still rich and important regional wetland. Figure 9-1 indicates that starting at any point within the environmental-biological interaction, the resource manager inevitably becomes involved in social, economic and legislative decisions which further limit his already constrained freedoms of action.

9.4 Local management

The day to day and year to year management of reserves involves continuous and often experimental decision-taking which ranges from the initiation of major projects such as altering soil or water levels to relatively minor matters including path and fence maintenance. Two publications of the British Trust for Conservation Volunteers (BROOKS, 1976; 1979) give detailed information concerning such work in wetland and coastal ecosystems. Management of the conflicting demands of conservation, agriculture, local government and the public, demand public relations work which may be one or the most important contributions to success.

Managerial decisions with long-term consequences such as the raising of water tables or stemming of successional change, cannot be undertaken lightly and yet may be essential to the maintainance of diversity or the preservation of particular animal and plant species.

The problems which arise from uncontrolled succession and changing management practices may be illustrated by reference to a small valley mire reserve, Cleddon Bog near Monmouth, South Wales now controlled by Gwent County Council. An accidental record from the beginning of the century, a watercolour painting, shows the bog with open pools and *Sphagnum* moss carpets: some local people remember the *Sphagnum* being collected as surgical

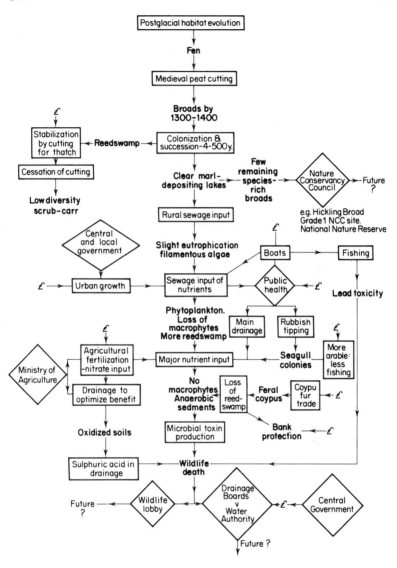

Fig. 9-1 Wetland development and management: an exercise in complexity. This simplified scheme shows the main factors influencing the present status of the Norfolk Broads in East Anglia and their associated wetland margins. Boxes represent causative changes, and bold type consequent conditions. The diamonds indicate legislative or consultative operations at a political level and the £ sign shows economic driving forces or consequences. (Sources mainly MOSS, 1980; ELLIS, 1965.)

dressings during the first World War, a use with prehistoric antecedents (GODWIN, 1978).

The catchment was afforested with pine, spruce and larch some 50–60 years ago and increased interception of precipitation, evaporation from the canopy and improved site drainage have probably caused some drying of the bog. Protection from burning, originally a management practice, has allowed succession to an irregularly tussocked purple moor grass (*Molinia caerulea*) cover with almost entire loss of the species-rich *Sphagnum* pools and carpets (Fig. 9-2).

Some wet-heath plants have probably become extinct, for example the white beaked sedge (*Rhynchospora alba*) while others such as sundew (*Drosera rotundifolia*) are now infrequent. Many others, including bog asphodel (*Narthecium ossifragum*) and cranberry (*Vaccinium oxycoccos*) are much reduced in number.

In 1962 a fire-break was cleared between the bog and the adjacent conifer plantation, the *M. caerulea* being considered an early spring fire risk. The fire-break was created by rotary ploughing and during the subsequent decade its bare peat surface became colonized by very large numbers of *Drosera rotundifolia, Nathecium ossifragum, Erica tetralix* and other wet-soil plants together with carpet-forming Sphagna which cannot compete in the closed moor-grass community. Gradually, as *M. caerulea* re-invaded, the diversity was reduced again until plants such as *D. rotundifolia* disappeared. As an experimental prelude to more extensive soil-surface management, trial plots were cleared of moor-grass in 1980 and, by the following year, already contained seedlings of *D. rotundifolia*, heath rush (*Juncus squarrosus*) and other species (Fig. 9-2).

Invasion of bog and fen surface by scrub may be controlled with 2-4-D weedkiller in diesel-fuel applied to individual shrubs or by digging, or by winching-out. FITTER and SMITH (1979) record the germination of buried yellow sedge (*Carex lepidocarpa*) seed in a depression left by tree-winching at Askham Bog in Yorkshire and, at Woodwalton Fen, Cambridgeshire, both fen violet (*Viola stagnina*) and the rare pale woodrush (*Luzula pallescens*) have appeared in tree-clearance hollows on fen peat (DUFFEY and WATT, 1971). GODWIN (1978) suggests a similar reason for the reappearance after 115 years of the fen ragwort (*Senecio paludosus*) in East Anglia.

The manipulation of soil surface is more locally controllable than raising the water table and one of the most rewarding management operations is the digging of sloping-sided depressions reaching to the water table or below. These provide habitats for many plant and animal species with different soil wetness or open-water requirements.

Colonization by robust species such as *M. caerulea* or by succession to scrub cover has, in the past, been controlled by fire. The rather rare occurrence of the tiny, prostrate *cranberry* (*Vaccinium oxycoccus*) in valley bogs in South Wales is almost certainly related to the practice of spring burning *M. coerulea* to promote early growth of sheep pasture. The cranberry would be outcompeted if the moor grass growth was not checked in this way.

Fig.9-2 Management trial plot in lowland mire vegetation, Cleddon Bog Local Nature Reserve, Gwent County Council, South Wales. Growth of purple moor grass (*Molinia caerulea*) seen in the foreground and invasion by scrub have reduced vegetational diversity of the site. Vegetation and peat was stripped from the plot in 1980 and by summer 1981 it contained over 30 seedlings of sundew (*Drosera rotundifolia*), germinated from a soil seed bank together with *Sphagum* moss, bog asphodel (*Narthecium ossifragum*) and other competition-intolerant bog plants. The dead dwarf-shrub in the foreground is ling (*Calluna vulgaris*), killed by an accidental fire. The results of these plot trials will be used to propose future management plans.

Changing management such as the cessation of reed-cutting and prevention of fire have caused losses of habitat by succession and in some cases specific efforts have been made to conserve important species. A Dutch variety of the large copper butterfly (*Lycaena dispar*) extinct in Britain since about 1851 was introduced to the East Anglian reserve Wood Walton Fen, in 1927 and survived until 1970 when a re-introduction was necessary. The introduced population has been supported by the provision of a number of saucer-shaped depressions of different age, planted with the foodplant, great water dock (*Rumex hydrolapathum*) (Duffey in DUFFEY and WATT 1971). Another fenland butterfly, the swallow-tail (*Papilio machaon*) is extinct at the oldest fenland reserve, Wicken Fen, and should it be re-introduced attempts will have to be made to culture the milk parsley (*Peucadanam palustre*) which has become reduced in population by loss of its mixed-sedge habitat.

Further sources of reference are scattered through a very large literature but the following publications may be useful. DUFFEY and WATT (1971) and GOOD *et al.* (1978) contain papers on wetland management while NEWBOLD (1977) outlines the relationship between wetland conservation and agriculture. Recent volumes of journals such as *Biological Conservation* contain papers ranging from discussion of particular plant and animal species to regional management problems.

9.5 Artificial wetlands

The need for irrigation water and hydro-electricity has, in the last 50 years, created new wetland areas. Many of the consequences were unforeseen by engineers though predicted by ecologists and affect both human and animal health as well as the future survival of some animal and plant species.

The main impact on human health is through waterborne disease, affecting many hundred millions of people; malaria alone causes one to two million deaths each year. The most severe problems are snail-vectored flatworms diseases such as schistosomiasis (bilharzia), mosquito-borne protozoal, viral and nematodal infections (malaria, arboviral diseases and filariasis) and bacterial leptospirosis which is increasing wherever irrigation is accompanied by large populations of rodents.

Schistosoma infection is caused by the penetration of the trematode cerceria through human skin which has contracted water containing the intermediate host-snail. Some 69% of children on the shores of the artificial Lake Kariba are infected with *S. haematobium* which is carried by *Bulinus* spp. feeding on shallow water beds of hornwort (*Ceratophyllum demersum*). The originally fast-flowing water of the Volta river could not support such weedbeds. The Aswan dam, on the Nile, likewise increased *S. mansonia* infection from two to 75% of the population (MOSS, 1980).

Filarial nematodes are carried by mosquitoes, for example, *Brugia malayi* by *Mansonia* spp., the larvae of which tap the intercellular airspaces of hydrophytes (section 4.1) and have been much increased by the growths of plants such as water hyacinth (*Eichhornia crassipes*) and water lettuce (*Pistia stratiotes*) which have spread immensely in impounded tropical rivers.

Mosquitoes are also the vectors for many arbovirus diseases including yellow fever, and for malaria caused by three important protozooan *Plasmodium* spp.

Again the effects of both impoundment and distribution of irrigation water have been the increase of the vector mosquitoes.

The liver fluke worms of sheep and other ruminants belonging to the genera *Fasciola* and *Fascioloides* are intermediately hosted by shallow water and wet vegetation snails of the genus *Lymnaea*. Many of the British pastures formerly managed as water meadows must have been particularly susceptible to fluke infestation and similarly the shore waters of impoundments and irrigation canals have become sources of infection.

The first attempts at vector control, for example asphyxiation of mosquito larvae with oil films, were rapidly followed by full-scale chemical warfare as industry provided new molluscicides and insecticides. The third generation of control measures must be more sophisticated and less generally damaging to the environment. Sound planning of water distribution and drainage schemes together with health education may provide almost total control with little need for pesticides but such developments may be many years away and often hindered by industial and governmental policies.

Wetland water supplies are often carried by rivers. The construction of impoundments eliminates seasonal flooding and removes the pulse stabilization which may have prevented successional infilling. It also prevents the transfer of nutrient-containing silt, so changing both the ecology and agriculture of the flood plain as well as depriving fisheries in nearshore waters of a detrital food supply. The Mediterranean fishery off the Nile delta has declined since its food-source was cut-off by the Aswan High Dam and the delta now needs artifical fertilizers (MOSS, 1980). It is not known whether the yield of the newly-created inland fishery will balance this economic loss.

10 Project Work

10.1 Fieldwork

Many small areas of wetland remain undetected and undescribed. Records of their biota are valuable and of use to local and national conservational bodies. Isolated small wetlands provide sites in which the distribution of animals and plants in relation to soil wetness may easily be studied. General ecological methods are described by CHAPMAN (1976) and extraction of soil organisms by JACKSON and RAW (1966).

Small wetlands surrounded by agricultural land receive farm drainage and are interfered with by cattle grazing. The effects of these factors in altering successional change may be studied, very often changes of land management such as fencing and draining.

The breakdown of litter in wetlands, the first stage in its mineralization is not well understood. Studies of the animals concerned and the rates of dry weight loss may be made either by sampling natural litter or using net bags in which portions of litter may be exposed (JACKSON and RAW, 1966).

An enormous amount remains to be discovered concerning animal life in wetlands and the factors which govern habitat preference and niche utilization. Field studies of populations and their behaviour may be supported by laboratory work with arthropods, molluscs and other invertebrates to establish food preferences and optimum environmental conditions.

Soil wetness may be measured by returning samples to the laboratory for dry weight determination. Use specimen tubes or small sealed polythene bags to avoid drying in transit. Water-table level may be monitored using an auger hole in which is inserted a length of plastic tube to avoid collapse (garden hose perforated at intervals).

Soil pH and redox potential may be measured in the field using a portable pH meter with, respectively, a glass-calomel soil pH electrode and a platinum-calomel electrode pair (ALLEN, 1974). Useful spot tests of soil reduction are those for Fe^{2+} and H_2S. For the former a portion of freshly extracted soil is pressed against a filter paper and the reverse side of the moist spot tested with potassium ferricyanide solution. A blue coloration is positive for Fe^{2+}. Sulphide may be detected by acidifying a soil sample with 10% HC1: evolution of H_2S gives a 'rotten egg' odour and the gas blackens filter paper impregnated with lead acetate.

10.2 Laboratory: chemical analysis

Conventional techniques are used (ALLEN, 1974) but if oxidizable substances such as Fe^{2+}, Mn^{2+} and H_2S are to be measured, either in solution or as part of the exchangeable cation complex, the soil samples must be taken directly into anaerobic (boiled) water or other extractant such as ammonium acetate and the container sealed without trapping air. Analyses should be undertaken quickly or toluene added to prevent further microbial action. Seasonal or tidal pulse changes of reducible elements are particularly interesting as are their correlations with the distribution of plants and animals.

10.3 Laboratory: microscopic study

The ecology of animals and micro-organisms within the soil micro-environment, in particular with relation to root radial oxygen loss has hardly been studied at all. Possible techniques are the microscopic study of thin sections cut after stabilization of soil cores or aggregates with either epoxy-resin or gelatine, direct microscopy of fractured soil surfaces and, for microorganisms a variety of culture techiques such as the buried slide, dilution plates and other approaches. Information may be sought from JACKSON and RAW (1966), PHILLIPSON (1970) and ANDREWS (1973).

Pollen analysis of peat is time-consuming and critical identification of species is difficult. General variations are, however, distinguishable and some important genera, for example *Pinus* in Britain, are easy to identify. Pollen may be concentrated into a pellet by centrifuging. MOORE and WEBB (1978) give extensive details, an identification key and photographs.

10.4 Laboratory: experimental studies

Waterlogging is one of the simplest soil characteristics to simulate in experimental conditions, for example by immersing pots of soil in a plastic bucket of water. Studies of species responses to waterlogging are thus quite easy and according to facilities, simple measurements of dry weight or photosynthetic gas exchange may be made coupled with monitoring of soil redox and chemical conditions. Short term studies of changes after flooding such as development of leaf epinasty are also possible.

Root anatomy varies according to degree of waterlogging. Comparison of species response to different waterlogging regimes or depth of water table is possible or roots may be grown in anaerobic nutrient media for such studies. Root radial oxygen loss is a function of root porosity and may be measured if a suitable oxygen electrode is available or a suitable redox dye (indigo carmine reduced with sodium dithionite) may be used as a semi-quantitative measure of oxygen loss (Fig. 3-4).

Further Reading and References

ALLEN, S. (1974). *Chemical Analysis of Ecological Materials*. Blackwell, Oxford.

ANDREWS, W.A. (1973). *Soil Ecology*. Prentice Hall, New Jersey.

ARMSTRONG, W. (1980). Aeration in higher plants. *Adv. bot. Res.*, **7**, 225–332.

AXELL, B. (1977). *Minismere: Portrait of a Bird Reserve*. Hutchinson, London.

BARNES, R.S.K. (1979). *Coasts & Estuaries*. Hodder & Stoughton, London.

BARNES, R.S.K. and MANN, K.H. (1980). *Fundamentals of aquatic ecosystems*, Blackwell, Oxford.

BEADLE, L.C. (1981). *The Inland Waters of Tropical Africa*. 2nd ed. Longman, London.

BORGIOLI, A. and CAPPELLI, G. (1979). *The Living Swamp*. Orbis, London.

BRIDGES, E.M. (1978). *World Soils*. 2nd edition. Cambridge University Press, Cambridge.

BROOKS, A. (1976). *Waterways and Wetlands*. British Trust for Conservation Volunteers, London.

BROOKS, A. (1979). *Coastlands*. British Trust for Conservation Volunteers, London.

BURTON, L.D. and LISS, P.S. (1976). *Estuarine Chemistry*. Academic Press, New York.

CARR, A. (1973). *Florida's Everglades*. Time-Life Books, Amsterdam.

CHAPMAN, S.E. (1976). *Methods in Plant Ecology*. Blackwell, Oxford.

CHAPMAN, V.J. (1977). *Wet Coastal Ecosystems*. Elsevier, Amsterdam.

CRAWFORD, R.M.M. (1976). Tolerance of anoxia and the regulation of glycolysis in tree roots. In M.G.R. Connel and F.T. Last, *Tree Physiology and Yield Improvement*. Academic Press, London, p.p. 387–401.

DIMBLEBEY, G.W. (1977). *Ecology and Archaeology*. Studies in Biology no. 77). Edward Arnold, London.

DREW, M.C., JACKSON, M.B. and GIFFORD, S. (1979). Ethylene-promoted adventitious rooting and development of cortical air spaces (aerenchyma) in roots may be adaptive responses to flooding in *Zea mays* (L.). *Planta*, **147**, 83–8.

DUFFEY, E. and WATT, A.S. (1971). *The Scientific Management of Animal and Plant Communities for Conservation*. Blackwell, Oxford.

DYKYJOVA, D. & KVÉT, J. (1978). *Pond Littoral Ecosystems*. Springer, Berlin.

ELLIS, E.A. (1965). *The Broads*. Collins, London.

ETHERINGTON, J.R. (1982). *Environment and Plant Ecology*, 2nd edition. Wiley, Chichester.

FITTER, A.H. and SMITH, W. (1979). *A Wood in Ascam: A study in Wetland Conservation*. Ebor Press, York.

GLOB, P.V. (1969). *The Bog People*. Cornell, U.P. Cornell.

GODWIN, H. (1978). *Fenland*. Cambridge University Press, Cambridge.

GODWIN, H. (1981). *The Archives of the Peat Bogs*. Cambridge University Press, Cambridge.

GOOD, R.E., WHIGHAM, D.F. and SIMPSON, R.C. (1978). *Freshwater Wetlands*. Academic Press, New York.

HASLER, A.D. (1975). *Coupling of Land & Water Systems*. Springer, Berlin.

IUCN–UNEP (1980). *A Directory of Western Palearctic Wetlands*. IUCN–UNEP, Gland.

IUCN–UNEP–WWF (1980). *World Conservation Strategy*. IUCN. Gland.

JACKSON, R.M. and RAW, F. (1966). *Life in the Soil*. Studies in Biology no. 2. Edward Arnold, London.

JONES, H.E. and ETHERINGTON, J.R. (1970). Comparative studies of plant growth and distribution in relation to waterlogging. I. The survival of *Erica cinerea* L. and *Erica tetralix* L. and its apparent relationship to iron and manganese uptake in waterlogged soil. *J. Ecol.*, **58**, 487–96.

KEELEY, J.E. (1979). Population differentiation along a flood frequency gradient: physical adaptations to floods in *Nyssa sylvatica. Ecol. Monogr.*, **49**, 89–108.

KUHNELT, W. (1961). *Soil Biology*. Faber and Faber, London.

LAING, H.E. (1940). Respiration of the rhizomes of *Nuphar advenum* and other water plants. *Amer. J. Bot.*, **27**, 574–81.

LEITH, H. and WHITTAKER, R.H. (1975). *Primary Productivity in the Biosphere*. Springer, Berlin.

McLUSKY, D. (1981). *The Estuarine Ecosystem*. Blackie, Glasgow.

MANN, K.H. (1982). *Ecology of Coastal Waters*. Blackwell, Oxford.

MOORE, P.D. and BELLAMY, D.J. (1973). *Peatlands*. Elek Science, London.

MOORE, P.D. and WEBB, J.A. (1978). *An Illustrated Guide to Pollen Analysis*. Hodder and Stoughton, London.

MOSS, B. (1980). *Ecology of Freshwaters*. Blackwell, Oxford.

NEWBOLD, C. (1977). Wetlands and Agriculture. In: (Eds) J. Davidson and R. Lloyd. *Conservation and Agriculture*. Wiley, Chichester. p.p. 57–9.

PHILLIPSON, J. (1970). *Methods of Study in Soil Ecology*. UNESCO, Paris.

POMEROY, L.R. and WIEGERT, R.G. (1981). *The Ecology of a Salt Marsh*. Springer, Berlin.

RANWELL, D.S. (1972). *Ecology of Salt Marshes and Sand Dunes*. Chapman and Hall, London.

REID, G.K. and WOOD, R.D. (1976). *Ecology of Inland Waters & Estuaries*. 2nd ed. Von Nostrand, N.Y.

SLACK, A. (1979). *Carnivorous Plants*. Ebury Press, London.

SMITH, A.M. and ap REES, T. (1979). Pathways of carbohydrate fermentation in the roots of marsh plants. *Planta*, **146**, 327–34.

STERLING, T. (1973). *The Amazon*. Time-Life Books, Amsterdam.

TANSLEY, A.G. (1939). *The British Islands and their Vegetation*. Cambridge University Press, Cambridge.

WALKER, D. (1970). Direction and rate in some British post-glacial hydroseres. In D. Walker and R.G. West. *Studies in the Vegetational History of the British Isles*. Cambridge University Press, Cambridge, p.p. 117–39.

WALLWORK, J.A. (1976). *The Distribution and Diversity of Soil Fauna*. Academic Press, London.

WELCH, E.B. (1980). *The Ecological Effects of Waste Waters*. C.U.P., Cambridge.

WEST, R.G. (1968). *Pleistocene Geology and Biology*. Longman, London.

WHITTON, B.A. (1975). *River Ecology*. Blackwell, Oxford.

WHITTON, B.A. (1980). *Rivers, Lakes & Marshes*. Hodder & Stoughton, London.

Index